00-HHE-6

IDENTIFYING TECHNOLOGIES AND COMMUNICATING THE BENEFITS AND RISKS OF DISINFECTING WET WEATHER FLOWS

by:

Brown and Caldwell
Syracuse, NY

Syracuse Research Corporation
Syracuse, NY

Wet Weather Environmental Technology Company
Roswell, GA

2005

WERF

The Water Environment Research Foundation, a not-for-profit organization, funds and manages water quality research for its subscribers through a diverse public-private partnership between municipal utilities, corporations, academia, industry, and the federal government. WERF subscribers include municipal and regional water and wastewater utilities, industrial corporations, environmental engineering firms, and others that share a commitment to cost-effective water quality solutions. WERF is dedicated to advancing science and technology addressing water quality issues as they impact water resources, the atmosphere, the lands, and quality of life.

For more information, contact:
Water Environment Research Foundation
635 Slaters Lane, Suite 300
Alexandria, VA 22314-1177
Tel: (703) 684-2470
Fax: (703) 299-0742
www.werf.org
werf@werf.org

This report was co-published by the following organization. For non-subscriber sales information, contact:

IWA Publishing
Alliance House, 12 Caxton Street
London SW1H 0QS, United Kingdom
Tel: +44 (0) 20 7654 5500
Fax: +44 (0) 20 7654 5555
www.iwapublishing.com
publications@iwap.co.uk

Library of Congress Catalog Card Number: 2003116837

IWAP ISBN: 978-1-84339-709-0

Brown and Caldwell

The research on which this report is based was funded, in part, by the United States Environmental Protection Agency through Cooperative Agreement No. CR-827345-01 with the Water Environment Research Foundation (WERF). Unless an EPA logo appears on the cover, this report is a publication of WERF, not EPA. Funds awarded under the Cooperative Agreement cited above were not used for editorial services, reproduction, printing, or distribution.

ACKNOWLEDGMENTS

The project team would like to acknowledge and thank the Onondaga County Department of Water Environment Protection for allowing the use of their collection system and wastewater treatment plant for this project. The project team would also like to acknowledge and thank the local stakeholders who committed time and information to this project. All stakeholders contributed to the findings of this report. In addition, staff members at all the project team organizations made critical contributions to the work and are to be thanked for their outstanding support. Finally, this project was funded by the United States Environmental Protection Agency and administered by the Water Environment Research Foundation.

Report Preparation

Principal Investigator:
Peter E. Moffa, P.E.
Brown and Caldwell, Syracuse, NY

Co-Investigator:
John J. LaGorga
Brown and Caldwell, Syracuse, NY

Project Team:
Linda Henry, Ph.D.
Brown and Caldwell, Walnut Creek, CA

Phillip Goodrum, Ph.D.
Syracuse Research Corporation, Syracuse, NY

Mark Boner, P.E.
Wet Weather Environmental Technology Company, Roswell, GA

Francis Doherty, Ph.D.
AquaTox Research, Syracuse, NY

Thomas Alexander
O'Brien and Gere Laboratories, Syracuse, NY

Project Subcommittee

Sydney Munger, Chair
Parametrix, Inc., Bellevue, WA

Joyce Perdek Walling, P.E.
U.S. EPA, Edison, NJ

Richard J. Field, P.E.
U.S. EPA, Edison, NJ

Arthur Meyers, Ph.D.
Eastman Chemical Company, Kingsport, TN

Paula Kehoe
San Francisco Public Utilities Commission, San Francisco, CA

Jo Anne Shatkin, Ph.D.
The Cadmus Group, Watertown, MA

Local Stakeholder Contributors

Richard Elander, P.E.
Onondaga County Department of Water Environment Protection

Stephen Martin
Onondaga County Department of Water Environment Protection

Thomas Friedrich
Trout Unlimited

Michael Pacelli
Concerned Citizen Representative

Russell Nemachek, P.E.
Onondaga County Department of Health

Elizabeth Colye, P.E.
New York State Department of Environmental Conservation, Region 7

John Ferrante, Ph.D.
Central New York Regional Planning & Development Board

James Hassett, Ph.D.
State University of New York College of Environmental Science and Forestry

Water Environment Research Foundation Staff

Director of Research: Daniel M. Woltering, Ph.D.
Project Managers: Dean Carpenter, D.E.E.
Christy Bixler, P.E.

ABSTRACT AND BENEFITS

Abstract:

The primary objective of this report is to identify and communicate the benefits and risks of disinfecting wet weather flows by evaluating the effectiveness of available disinfection technologies and identifying disinfection by-products and their potential risks to aquatic and human life. A decision-making framework is presented that could be used as a model to guide combined sewer overflow, sanitary sewer overflow and stormwater disinfection control policies.

This project was implemented as a case study in collaboration with the Onondaga County Department of Water Environment Protection (Syracuse, NY). A literature review was completed that identified available disinfection technologies and the associated disinfection by-products. A disinfection demonstration was conducted to verify findings of the literature review and fill data gaps. The data collected during this project were used by local stakeholders to select the most appropriate technologies. Information regarding the benefits and risks of disinfecting wet weather flows was also presented to the public during a public workshop.

All four technologies that were piloted, namely chlorination/dechlorination, chlorine dioxide (ClO_2), ozonation, and ultraviolet light (UV), achieved the United States Environmental Protection Agency's bacterial criteria. Ozonation and UV produced no measurable residuals or by-products and resulted in no effluent aquatic toxicity as measured through whole effluent toxicity (WET) testing. Chlorination/dechlorination and chlorine dioxide did produce residuals and disinfection by-products including chlorate, chlorite, trihalomethanes, and haloacetic acids. With the exception of chlorite, these by-product concentrations were less than drinking water standards. Chlorination/dechlorination and chlorine dioxide did result in effluent aquatic toxicity as measured through WET testing. In the case of chlorination/dechlorination, the toxicity appeared to be related to residual chlorine, and in the case of chlorine dioxide, the toxicity appeared to be related to residual chlorine dioxide and/or chlorite. Residuals can be controlled through proper quenching or reducing agents in a full-scale application. The chlorite in the chlorine dioxide effluent was a by-product of the particular chlorine dioxide production method; this could be avoided by a generation method that produces a chlorite-free product, which would be more appropriate for a full-scale application.

Based on findings from the literature review and disinfection demonstration, hazards associated with wet weather flow disinfection appear low, but the potential for public opposition can be high. Therefore, proper communication planning to include stakeholders in the decision-making process and conduct public outreach is required. Stakeholders were included throughout the project through workshops.

Benefits:

- Develops and implements a decision-making framework to assist communities in making appropriate decisions regarding disinfecting wet weather flows.
- Provides a literature review pertaining to the disinfection of wet weather flows and associated residual and disinfection by-products, including a hazard identification review.

- Provides a summary of a disinfection demonstration test plan, which subscribers can use to design their own disinfection demonstrations, as it may be necessary for municipalities to perform site-specific disinfection demonstrations to fill data gaps.
- Provides results and conclusions from nine disinfection demonstration events. The results and conclusions can be used in the decision-making framework.
- Develops and implements a risk communication plan. Utilities can use this information to develop their own communication based on lessons learned from this project.

Keywords: wet weather flow, combined sewer overflow (CSO), sanitary sewer overflow (SSO), stormwater (SW), disinfection, pathogens, disinfection by-products (DBPs), risk assessment, risk communication

TABLE OF CONTENTS

Acknowledgments iii
Abstract and Benefits v
List of Tables x
List of Figures xii
List of Acronyms xiii
Executive Summary ES-1
Chapter 1.0 Introduction 1-1
1.1 Rationale for Wet Weather Flow Disinfection 1-1
1.1.1 Stormwater 1-1
1.1.2 Combined Sewer Overflows 1-1
1.1.3 Sanitary Sewer Overflows 1-1
1.1.4 Wet Weather Flow Disinfection 1-2
1.2 Objectives and Scope of Project 1-3

Chapter 2.0 Decision-Making Framework 2-1
2.1 Risk Management 2-1
2.2 Decision-Making Framework 2-1
2.2.1 Information Inputs 2-2
2.2.2 Benefit/Risk/Feasibility Assessment 2-2
2.2.3 Technology or Policy Selection 2-3
2.2.4 Stakeholder Involvement 2-3

Chapter 3.0 Literature Review 3-1
3.1 Wet Weather Flow Disinfection Technology 3-1
3.2 Benefits of Disinfecting Wet Weather Flows 3-2
3.3 Risks of Disinfecting Wet Weather Flows—Hazard Identification 3-3
3.3.1 Hazard Identification 3-4
3.4 Bacterial Standards 3-7
3.5 Full-Scale Facilities 3-8

Chapter 4.0 Disinfection Demonstration 4-1
4.1 Disinfection Demonstration Scope of Work 4-1
4.2 Disinfection Demonstration Methods and Materials 4-2
4.2.1 Disinfection Equipment 4-2
4.3 Disinfection Demonstration Results 4-3
4.3.1 Event Summary 4-3
4.3.2 Pilot Influent 4-5
4.3.3 Chlorination 4-9
4.3.4 Chlorine Dioxide 4-20
4.3.5 Ozonation 4-31
4.3.6 Ultraviolet Light 4-36
4.3.7 Dose-Response Relationships 4-37

Chapter 5.0 Applications for Full-Scale Facilities 5-1
5.1 Site-Specific Application of Findings 5-1
5.2 Application of Findings to Onondaga County Case Study 5-2
5.2.1 Bacterial Criterion 5-2
5.2.2 Water Quality Adjustments 5-2
5.2.3 Receiving Water Dilution Adjustments 5-3
5.2.4 Estimated Environmental Concentrations 5-6
5.3 Screening-level Risk Assessment 5-7
5.3.1 General Approach 5-7
5.3.2 Hazard Identification 5-8
5.3.3 Exposure Assessment 5-9
5.3.4 Toxicity Assessment 5-15
5.3.5 Risk Characterization 5-18
5.3.6 Screening-level Risk Assessment Conclusions 5-22
5.4 Operations 5-24
5.4.1 Developing a Disinfection Dose Feed Program 5-24
5.5 Conceptual Costs 5-28
5.5.1 Storm Event and Annual Volumes 5-28
5.5.2 Chlorination/Dechlorination 5-29
5.5.3 Chlorine Dioxide 5-31
5.5.4 Ozonation 5-33
5.5.5 Ultraviolet Light 5-35
5.5.6 Conceptual Cost Summary 5-37

Chapter 6.0 Communication 6-1
6.1 Decision-Making Framework Communication Plan 6-1
6.2 Communicating Tools 6-2
6.3 Communicating Risk 6-3
6.4 Lessons Learned 6-4

Chapter 7.0 Conclusions 7-1

Chapter 8.0 Hazard Identification Literature Review 8-1
8.1 Hazard Identification Summary 8-1
8.2 Hazard Identification Databases 8-1
8.2.1 HSDB—Hazardous Substances Data Bank 8-1
8.2.2 IARC Monographs—International Agency for Research on Cancer 8-1
8.2.3 IRIS—Integrated Risk Information System 8-1
8.2.4 ECOTOX/AQUIRE 8-2
8.3 Chlorine Residuals 8-2
8.3.1 Mammalian (Human) Toxicity 8-2
8.3.2 Aquatic Ecotoxicity 8-4
8.4 Chlorine Dioxide, Chlorite, and Chlorate 8-4
8.4.1 Mammalian (Human) Toxicity 8-4
8.4.2 Aquatic Ecotoxicity 8-6
8.5 THM: Chloroform 8-7
8.5.1 Mammalian Toxicity 8-7

8.5.2 Aquatic Ecotoxicity 8-8
8.6 THM: Bromodichloromethane........ 8-9
8.6.1 Mammalian Toxicity........ 8-9
8.6.2 Aquatic Ecotoxicity 8-10
8.7 THM: Dibromochloromethane 8-10
8.7.1 Mammalian Toxicity........ 8-10
8.7.2 Aquatic Ecotoxicity 8-11
8.8 THM: Bromoform 8-12
8.8.1 Mammalian Toxicity........ 8-12
8.8.2 Aquatic Ecotoxicity 8-13
8.9 HAA: Monochloroacetic Acid 8-13
8.9.1 Mammalian Toxicity........ 8-13
8.9.2 Aquatic Ecotoxicity 8-14
8.10 HAA: Dichloroacetic Acid 8-14
8.10.1 Mammalian Toxicity........ 8-14
8.10.2 Aquatic Ecotoxicity 8-19
8.11 HAA: Trichloroacetic Acid 8-20
8.11.1 Mammalian Toxicity........ 8-20
8.11.2 Aquatic Ecotoxicity 8-20
8.12 HAA: Monobromoacetic Acid 8-21
8.12.1 Mammalian Toxicity........ 8-21
8.12.2 Aquatic Ecotoxicity 8-21
8.13 HAA: Dibromoacetic Acid 8-21
8.13.1 Mammalian Toxicity........ 8-22
8.13.2 Aquatic Ecotoxicity 8-24

References........R-1

LIST OF TABLES

ES-1 Draft U.S. EPA Criteria for Maximum Bacterial Indicator Organisms in Fresh-Recreational Waters....ES-3
ES-2 Disinfectant Doses Required to Meet Bacterial Criteria....ES-5
ES-3 Site-Specific Doses and Resulting Residuals and DBPs in the Effluent....ES-9
ES-4 Estimated Environmental Concentrations from the Chlorination Process....ES-10
ES-5 Estimated Environmental Concentrations from the Chlorine Dioxide Process....ES-10
3-1 Disinfection Residuals and DBPs....3-5
3-2 Disinfection Residual and DBP Hazard Identification Summary....3-6
3-3 Draft U.S. EPA Indicator Bacterial Organisms Criteria....3-7
4-1 Disinfection Demonstration Sampling Event Statistics....4-3
4-2 Disinfection Demonstration Influent Water Quality Data....4-7
4-3 Disinfection Demonstration Influent Water Quality R^2 Comparison....4-7
4-4 Bacterial Data Summary for Chlorination Disinfection....4-9
4-5 Chlorine Dose Required to Meet Bacterial Criteria....4-12
4-6 Chlorination/Dechlorination Effluent Disinfection Residual and DBP Data....4-13
4-7 Chlorination Residuals and DBPs for Typical Wet Weather Flow at METRO....4-16
4-8 Toxicity Literature Review Findings for Chlorination....4-17
4-9 Bacterial Data Summary for Chlorine Dioxide Disinfection....4-20
4-10 Chlorine Dioxide Dose Required to Meet Bacterial Criteria....4-21
4-11 Chlorine Dioxide Effluent Disinfection Residual and DBP Data....4-24
4-12 Chlorine Dioxide Residual and DBP for Typical Wet Weather Flow at METRO....4-27
4-13 Toxicity Literature Review Findings for Chlorine Dioxide....4-31
4-14 Bacterial Data Summary for Ozonation Disinfection....4-31
4-15 Ozone Doses Required to Meet Bacterial Criteria....4-33
4-16 Ozonation Effluent Disinfection Residual and DBPs Data....4-36
4-17 Bacterial Data Summary for UV Disinfection....4-37
4-18 UV Dose Required to Meet Bacterial Criteria....4-39
4-19 UV Effluent Disinfection Residual and DBPs Data....4-39
5-1 Targeted Dose and Resulting Residuals and DBPs in the Disinfected Effluent....5-6
5-2 Targeted Dose and Resulting Residuals and DBPs in Onondaga Creek....5-7
5-3 Targeted Dose and Resulting Residuals and DBPs in Onondaga Lake....5-7
5-4 Volatilization Half-life in a Model Stream....5-12
5-5 Input Parameter Values Used to Calculate CDI....5-13
5-6 Chronic Daily Intake Results (mg/kg-day) by Exposure Pathway, for Noncancer Risk Calculation....5-14
5-7 Chronic Daily Intake Results (mg/kg-day) by Exposure Pathway, for Cancer Risk Calculation....5-14
5-8 Toxicity Reference Values Available for Chemicals with EECs....5-17
5-9 Toxicity Reference Values Available for Chemicals with EECs....5-21
5-10 Hazard Identification Literature Review Findings....5-23
5-11 Hazard Quotient for Chlorination and Chlorine Dioxide Residuals and By-products....5-24

5-12 Chlorination/Dechlorination Conceptual Capital Costs and Chemical and Energy Costs. 5-30
5-13 Chlorine Dioxide Conceptual Capital Costs and Chemical and Energy Costs. 5-32
5-14 Ozonation Conceptual Capital Costs and Chemical and Energy Costs. 5-34
5-15 Ultraviolet Light Conceptual Capital Costs and Chemical and Energy Costs. 5-36

LIST OF FIGURES

2-1 Decision-Making Framework 2-5
3-1 Types of CSO Facilities in 1998 3-8
4-1 Disinfection Demonstration Sampling Approach 4-4
4-2 Comparison of the Indicator Bacterial Organisms in the Influent. 4-8
4-3 Example of Chlorine Dose Normalization Process for Fecal Coliform 4-10
4-4 Chlorine Dose-Response Curves Normalized by COD/Temperature. 4-11
4-5 Comparison of the Indicator Bacterial Organisms in the Chlorine Effluent 4-14
4-6 Normalized Chlorine Dose and Pre-Dechlorination Effluent TRC. 4-18
4-7 Normalized Post-Dechlorination Effluent TRC and Effluent Toxicity. 4-18
4-8 Chlorine Dose and Effluent Chlorate 4-18
4-9 Chlorine Dose and Effluent Chloroform 4-19
4-10 Normalized Chlorine Dose and Effluent Dichloroacetic Acid 4-19
4-11 Normalized Chlorine Dose and Effluent Trichloroacetic Acid 4-19
4-12 Chlorine Dioxide Dose-Response Curves Normalized by COD/Temperature 4-22
4-13 Comparison of the Bacterial in the Chlorine Dioxide Effluent. 4-23
4-14 Chlorine Dioxide Dose and Pre-Reducing Agent Effluent Residual Chlorine Dioxide 4-27
4-15 Post-Reducing Agent Effluent Chlorine Dioxide Residual and Toxicity. 4-28
4-16 Chlorine Dioxide Dose and Effluent Chlorate 4-28
4-17 Chlorine Dioxide Dose and Pre-Reducing Agent Effluent Chlorite 4-28
4-18 Post-Reducing Agent Effluent Chlorite and Toxicity 4-29
4-19 Chlorine Dioxide Dose and Effluent Monochloroacetic Acid 4-29
4-20 Chlorine Dioxide Dose and Effluent Dichloroacetic Acid. 4-29
4-21 Chlorine Dioxide Dose and Effluent Trichloroacetic Acid 4-30
4-22 Ozone Dose-Response Curves Normalized by COD/Temperature 4-34
4-23 Comparison of the Bacterial in the Ozone Effluent 4-35
4-24 UV Dose-Response Curves Normalized by COD/Temperature 4-38
4-25 Comparison of the Bacterial in the UV Effluent 4-41
5-1 Chlorine Dose-Response Curves Normalized by COD/Temperature. 5-3
5-2 Wet Weather Flow Rate versus TSS Concentration, Site Specific Discharge 5-26
5-3 Chlorine Dose-Response Curves Normalized by TSS and Temperature. 5-26
5-4 Chlorine Dosing Schedule for Site Specific Discharge. 5-26
5-5 Chlorine Dioxide Dosing Schedule for Site Specific Discharge. 5-27
5-6 Ozone Dosing Schedule for Site Specific Discharge 5-27
5-7 UV Dosing Schedule for Site Specific Discharge 5-27
5-8 Conceptual Capital Costs and Chemical and Energy Costs 5-38

LIST OF ACRONYMS

AQUIRE	AQUatic toxicity Information REtrieval
AWQC	Ambient Water Quality Criteria
ATSDR	Agency for Toxic Substances and Disease Registry
BOD	Biochemical Oxygen Demand
CFU	Colony-Forming Unit
CSF	Cancer Slope Factor
COPC	Chemicals of Potential Concern
CDI	Chronic Daily Intake
CFS	Cubic Feet per Second
COD	Chemical Oxygen Demand
CSO	Combined Sewer Overflow
DBP	Disinfection By-Product
DCA	Dichloroacetic Acid
E-Beam	Electron Beam Irradiation
EC_{50}	Median Effective Concentration
ECOTOX	EPA Ecotoxicity Database
EEC	Estimated Environmental Concentration
ENU	Ethylnitrosourea
EEC	Expected Environmental Concentration
GPM	Gallons per Minute
HAA	Haloacetic Acids
HDPE	High-Density Polyethylene
HQ	Hazard Quotient
HSDB	Hazardous Substances Data Bank
I&I	Infiltration and Inflow
IARC	International Agency for Research on Cancer
IRIS	Integrated Risk Information System
LC_{50}	Median Lethal Concentration
LOAEL	Lowest Observed Adverse Effect Level
LOEC	Lowest Observed Effect Concentration
MA7CD10	Maximum Average Seven-Consecutive-Day Low Flow with a 10-Year Recurrence Interval
MATC	Maximum Allowable/Acceptable Toxicant Concentration
MCL	Maximum Contaminant Level
MCLG	Maximum Contaminant Level Goal
MGD	Million Gallons per Day
MNU	Methylnitrosourea
MRDLG	Maximum Residual Disinfection Level Goal
METRO	Metropolitan Sewage Treatment Plant
NAS	National Academy of Science
NCI	National Cancer Institute
NIEHS	National Institute of Environmental Health Sciences

NLM	National Library of Medicine
NOAEL	No Observed Adverse Effect Level
NOEC	No Observed Effect Concentration
NOEL	No Observed Effect Level
NPDES	National Pollutant Discharge Elimination System
NRC	National Research Council
NTP	National Toxicology Program
PHYTOTOX	EPA Terrestrial Plant Toxicity Database
ppb	Parts per Billion
ppm	Parts per Million
RfC	Reference Concentration
RfD	Reference Dose
RBC	Risk-Based Concentration
SpC	Specific Conductivity
SDWA	Safe Drinking Water Act
SSO	Sanitary Sewer Overflow
SW	Stormwater
TERRETOX	EPA Wildlife Toxicity Database
TOC	Total Organic Carbon
TOXNET	Toxicology Data Network
THM	Trihalomethane
TKN	Total Kjeldahl Nitrogen
TRC	Total Residual Chlorine
TSS	Total Suspended Solids
UF	Uncertainty Factor
UV	Ultraviolet
VSS	Volatile Suspended Solids
WET	Whole Effluent Toxicity
WWETCO	Wet Weather Engineering & Technology Company
WWTP	Wastewater Treatment Plant
WWF	Wet Weather Flow

EXECUTIVE SUMMARY

ES.1 Project Objectives

The primary objective of this report is to identify and communicate the benefits and risks of disinfecting wet weather flows with chlorine and alternative disinfectants, including chlorine dioxide (ClO_2), ozonation, and ultraviolet light (UV). The benefits of disinfection include the reduction of pathogens discharged to receiving waters that may be used for recreation or municipal drinking water supplies. The risks of disinfection include an increase in disinfection residuals and disinfection by-products (DBPs) to these waters.

ES.2 Research Approach

The three major phases of this project were:

1. *Literature Review*: Gather and summarize existing information on the effectiveness (i.e., benefits) of wet weather flow disinfection and the associated DBPs and their potential risks to human and aquatic life.
2. *Disinfection Demonstration*: Perform a disinfection demonstration to evaluate the effectiveness of disinfection technologies and identify associated DBPs and their potential risks to human and aquatic life.
3. *Communication*: Communicate the findings of the literature review and disinfection demonstration to stakeholders and the public.

The three phases of this project were implemented in collaboration with Onondaga County Department of Water Environment Protection and local stakeholders. A decision-making framework was developed and implemented for the purpose of selecting the most appropriate disinfection technologies for wet weather flows.

ES.3 Literature Review Findings

Most wet weather flows contain high levels of bacteria that may pose health risks to the recreational users of receiving waters. Water quality impacts of combined sewer overflows (CSO), sanitary sewer overflows (SSO) and stormwater (SW) discharges are highly site-specific, temporally variable, and sporadic in occurrence. The practitioner must understand the dynamics of wet weather flows, their impact on receiving water, and the usability-related characteristics of the receiving water in order to fully assess the level of control to comply with appropriate water quality standards or criteria.

Chlorine has long been the most common disinfectant for drinking water, wastewater, and wet weather flows. More recently, alternative disinfectants to chlorine have gained recognition.

The alternative disinfectants identified in the U.S. EPA's *Combined Sewer Overflow Technology Fact Sheet, Alternative Disinfection Methods* (U.S. EPA, 1999a) are:

- chlorine dioxide
- ozonation
- UV radiation
- peracetic acid
- electron beam irradiation (E-beam).

Other disinfectants that may also be appropriate for wet weather flows are bromine and mixed oxidants.

The most important reason for disinfecting water is to inactivate pathogenic organisms and thus limit waterborne disease. The benefits of disinfecting drinking water, wastewater and wet weather flows has been well documented since early in the twentieth century and is best documented for drinking water applications. For example, since the introduction of filtration and disinfection at drinking water treatment plants in the United States, waterborne diseases such as typhoid and cholera have been virtually eliminated; in Niagara Falls, NY, between 1911 and 1915, the number of typhoid cases dropped from 185 deaths per 100,000 population to nearly zero following introduction of filtration and chlorination (White, 1999).

Ambient water quality standards for recreational and other beneficial uses include criteria for indicator bacterial organisms that relate to gastrointestinal illness rates. Fecal coliform, *Escherichia coli,* and *Enterococcus* spp. are typical indicator bacterial organisms; the criteria include both 30-day geometric mean and maximum-day values for different illness rates to protect different levels of full-body contact recreational uses. The latter two indicator organisms have been shown to correlate to illness rates and are proposed by the U.S. EPA for adoption by states and tribes in its May 2002 draft, *Implementation Guidance for Ambient Water Quality Criteria for Bacteria.*

In this guidance, illness rates that may be selected for a particular waterbody range from 8 illnesses per 1,000 people to 14 per 1,000 for fresh-recreational waters and 8 per 1,000 to 19 per 1,000 for marine-recreational waters. Each illness rate for the two types of receiving waters has a corresponding 30-day geometric mean criterion and a range of maximum values. Maximum day values relate to the confidence limits around the data set. Designated beach areas should have the highest protection and, therefore, the most conservative confidence limits. Infrequently-used, full-body contact waters can have less-conservative limits. In other words, as confidences increase, less of a safety factor is needed and, therefore, less conservative assumptions can be implemented. Table ES-1 presents the proposed maximum bacterial criteria for 14 illnesses per 1,000 people in fresh-recreational waters.

Table ES-1. Draft U.S. EPA Criteria for Maximum Bacterial Indicator Organisms in Fresh-Recreational Waters. (14 illnesses per 1,000 people)

Indicator Organism	Maximum Value
Fecal coliform	4,274 CFU*/100 mL‡
E. coli	2,507 CFU/100 mL
Enterococcus spp.	656 CFU/100 mL

* colony-forming unit

‡ The maximum value for fecal coliform is not specified by the U.S. EPA. It was calculated using data collected from this disinfection demonstration. A relation between *E. coli* and fecal coliform was developed using the ratio of these organisms measured in the disinfected effluent. This relationship was then used to calculate the fecal coliform concentration that corresponded to the *E. coli* criterion of 2,507 CFU/100 mL, which was 4,274 CFU/100 mL.

Although chlorine disinfection has significantly decreased the incidence of waterborne disease in the United States for many years, growing concerns about chlorine's safety and effects on the environment have prompted communities to evaluate other disinfection methods (Hass, 1986). The concerns regarding chlorine disinfection are mainly in two areas: (1) safety in handling and use, and (2) disinfection residuals and DBPs created by the addition of chlorine to the wastewater.

ES.4 Disinfection Demonstration

Disinfection Demonstration Approach

Chlorination, chlorine dioxide, ozonation, and UV were deemed to be the most appropriate technologies for disinfection of wet weather flows and therefore were piloted for this disinfection demonstration. Peracetic acid and E-beam were not evaluated due to lack of full-scale experience. In the case of peracetic acid, on-site generation using two chemicals is needed, and this technology has not been implemented in a full-scale wet weather flow application. In the case of E-Beam, pretreatment requirements make it inappropriate for wet weather flow disinfection. Bromine was not evaluated separately from chlorine because bromination has similar disinfection residual and DBP concerns as chlorination.

The disinfection demonstration was a large bench-scale pilot test; each disinfection technology was sized to treat approximately 50–100 gallons of flow volume. The disinfection technologies were considered high-rate, meaning (1) the contact time for chlorination and chlorine dioxide was approximately 5 minutes, (2) the contact time for ozonation was approximately 3 minutes, and (3) the contact time for UV was approximately 5 seconds. The pilots were operated such that the disinfectant dose was varied for each demonstration event so that disinfectant dose versus bacterial reductions (dose-response) curves could be developed. The chlorination pilot included dechlorination with sodium bisulfite ($NaHSO_3$) for the purpose of reducing residual chlorine. After the first four events, ferrous sulfate was added as a reducing agent to the chlorine dioxide pilot. The addition this reducing agent was for the purpose of reducing chlorite concentrations in the pilot effluent, which were the result of chlorite contamination of the chlorine dioxide product.

The demonstration began in July of 2001, but because of a lack of rain, only four events were completed by October 2001. The demonstration was shut down for the winter months and reactivated in April of 2002. The remaining five demonstration events were completed between April and June of 2002.

Each wet weather event exhibited different flow and water quality characteristics. Flow was collected at the Metropolitan Sewage Treatment Plant (METRO), located in Syracuse, Onondaga County, NY, during wet weather periods. Sampling was done at the treatment plant rather than at a CSO discharge site to assure a more predictable and reliable wet weather flow. Flow was collected after screening and degritting but prior to primary clarification and was used as the influent to the disinfection pilot technologies. This water quality was consistent with preliminarily-treated combined sewage such as flows that have been treated by high-rate physical treatment devices (e.g., vortex technology). The events of autumn 2001 tended to have greater rain intensities than the spring 2002 events; however, the flow rates tended to be greater during the spring 2002 events. This was likely the result of greater infiltration due to high groundwater during the spring months.

Influent Water Quality

The raw influent water quality was characterized as a "weak" wastewater. The average biochemical oxygen demand (BOD) concentration was 76 mg/L, chemical oxygen demand (COD) averaged 182 mg/L, total organic carbon (TOC) averaged 64 mg/L, and total suspended solids (TSS) averaged 101 mg/L. Bacteria concentrations were typical of a CSO. Fecal coliform concentrations ranged from 1.0×10^5 to 3.5×10^7 CFU/100 mL, *E. coli* concentrations ranged from 3.0×10^4 to 2.2×10^7 CFU/100 mL, and *Enterococcus* spp. concentrations ranged from 2.9×10^3 to 7.2×10^6. The events were segregated by water temperature into three categories: four events had water temperatures between 12°C and 13°C); another four events had temperatures between 18°C and 22°C; and one event had a water temperature of 26°C. Temperature is significant because it strongly correlates with fecal coliform and *E. coli* concentrations (R^2 values of 0.81 and 0.71, respectively); as water temperatures increase, so do bacteria counts. The bacteria concentrations of the four cool-temperature events were generally one to two orders-of-magnitude less than the warmer temperature events.

Owing to the U.S. EPA criteria for *E. coli* and *Enterococcus* spp., an understanding of the relationship between the two indicator organisms and fecal coliform in the untreated and treated water is important. In the influent water, the correlation between fecal coliform and *E. coli* is strong, whereas the correlations with the other indicators are weak, showing only general trends. As such, these data indicate that at this sampling location, historical fecal coliform data can be transformed into *E. coli* data with some certainty. Such a transformation would not be appropriate from fecal coliform to *Enterococcus* spp. and between *E. coli* and *Enterococcus* spp.

Influent water was not toxic to fathead minnows for each of the first four demonstration events (whole effluent toxicity (WET) tests using the fathead minnow were only performed during the first four events). In each instance, there was no mortality observed among minnows exposed to whole influent sample for a period of 48 hours in preliminary screening tests. Influent water was also not toxic to the invertebrate *Ceriodaphnia dubia* for each of the nine demonstration events in this study. There was an insignificant level of mortality (<15%) among *C. dubia* exposed to whole influent (i.e., no dilution) after an exposure period of 48 hours in preliminary screening tests. Consequently, the reported median lethal concentration (LC_{50}) for each species for each collection event was >100%. Greater than 100% means an insignificant evel of mortality resulted from using an undiluted influent sample. Mictotox™ bacterial cultures exhibited minor reductions in luminescence on exposure to whole influent water in each of the nine collection events. Changes in luminescence in multi-replicate inhibition tests ranged from slight increases relative to control levels to no more than a 50% reduction in light output. Consequently, the reported 15-minute median effective concentrations (EC_{50}) for influent from each collection event was >100% (i.e., no dilution).

Effluent Bacterial Reductions

Effluent samples were collected from each of the disinfection technologies and compared to the influent samples for the purpose of developing disinfectant dose versus bacterial reduction (dose-response) curves. In addition to dose-response curves, disinfection residual and DBP production curves were developed for each of the disinfection technologies. These sets of curves were used to compare the bacterial reductions (i.e., benefits) and risks of the disinfection technologies.

The dose-response curves were originally developed using the applied doses and the resulting bacterial reductions, but the correlation between the doses and bacterial reductions was marginal. However, when the dose was normalized by water quality parameters such as temperature, COD, TSS, and total Kjeldahl nitrogen (TKN), the correlation markedly improved. The pH did not vary and therefore was not considered as a normalization parameter. The best correlations occurred when the applied doses were normalized by COD and temperature. The addition of other parameters, such as TSS and TKN, did not improve this relationship measurably. Therefore, the data analysis was performed with the COD/temperature normalized curves.

Using the normalized dose-response curves, the performance of the technologies was evaluated based on their ability to achieve the bacterial criteria presented in Table ES-1. Table ES-2 presents the disinfectant doses that were required to achieve the bacterial criteria for a water quality characterized by a COD of 200 mg/L and a temperature of 20°C.

Table ES-2. Disinfectant Doses Required to Meet Bacterial Criteria.*

Parameter	Chlorination	Chlorine Dioxide	Ozonation	UV
Fecal coliform	18 mg/L	6.3 mg/L	25 mg/L	110 mWsec/cm^2
E. coli	18 mg/L	6.6 mg/L	23 mg/L	100 mWsec/cm^2
Enterococcus spp.	22 mg/L	8.6 mg/L	20 mg/L	140 mWsec/cm^2

* Water quality characterized by COD of 200 mg/L and temperature of 20°C.

The chlorine dose required to meet the criteria for fecal coliform and *E. coli* was approximately the same but increased markedly to meet the *Enterococcus* spp. criterion. A similar trend was observed for chorine dioxide and UV. For chlorination, chlorine dioxide, and UV, between 20% and 36% higher doses were required to achieve the *Enterococcus* spp. criterion as compared to the fecal coliform and *E. coli* criterion. In contrast, ozonation was most effective at achieving the *Enterococcus* spp. criteria, requiring 22% less ozone as compared to the fecal coliform and *E. coli* criteria.

Approximately one-third the dosage of chlorine dioxide was required to achieve the same bacterial reductions as chlorination. In general, chlorine dioxide in water is 10 times more soluble, making it a broader bactericide and an effective viricide (White, 1999). In contrast, a higher ozone dose was required to achieve the same bacterial reductions as chlorination, which is contrary to the literature regarding ozonation (Morris, 1975). Research by J.C. Morris illustrates that ozone is a more powerful germicide against all classes of organisms by a factor of 10–100. This discrepancy may be the result of a combination of ozone demand and difficulties in general with measuring ozone residuals.

Effluent Disinfection Residuals, By-products, and Toxicity

Equally important as measuring and evaluating the performance of the disinfection technologies to inactivate bacteria was the development of disinfection residual and by-product curves for each of the disinfection technologies. In summary, chemical disinfection of wet weather flows with chlorine and chlorine dioxide resulted in disinfectant residuals and by-products, whereas ozonation and UV disinfection did not produce residuals or DBPs.

Chlorine and chlorine dioxide residuals were generally found to correlate to the applied dose before the addition of a reducing agent. When applying 18 mg/L of chlorine and 6.6 mg/L of chlorine dioxide (Table ES-2), the resulting residuals were 4.3 mg/L and 0.14 mg/L for chlorine and chlorine dioxide, respectively.

DBPs measured included chlorates and chlorites. The U.S. EPA Microbial and Disinfection By-Products Rule limits ambient concentrations for chlorite to 0.8 mg/L. There is no current limit for chlorate or studies indicating toxicity effects for this substance, but the U.S. EPA has suggested limitation of its production without specifying a maximum concentration.

The production of chlorate was found to be proportional to the applied dose in both chlorination and chlorine dioxide pilots. When applying 18 mg/L of chlorine and 6.6 mg/L of chlorine dioxide (Table ES-2), chlorate production was estimated to be 3.0 mg/L and 3.3 mg/L for chlorine and chlorine dioxide disinfection, respectively. For the same applied doses, chlorite production was estimated as zero and 18 mg/L for chlorination and chlorine dioxide, respectively. These chlorate and chlorite values are before the addition of the reducing agents. The chlorite concentration (18 mg/L) in the chlorine dioxide effluent exceeded the applied dose (6.6 mg/L) because there was residual chlorite in the chlorine dioxide product as a result of the generation method employed for this demonstration. Full-scale applications would employ generation processes that produce a chlorite-free product.

When studying the effect of reducing agents, sodium bisulfate successfully reduced the residual chlorine from the chlorination process in most instances. However, four events occurred during which an insufficient quantity of sodium bisulfite was added, and residual chlorine remained present in the effluent. During one of the four events, additional sodium bisulfite was added to ensure that the residual chlorine was below the detection limit. Ferrous sulfate was also successful at reducing both chlorine dioxide residual and chlorite concentration in events where the reducing agent was added in sufficient quantities. Neither reducing agent was found to affect chlorate concentrarions in the effluent.

Trihalomethanes (THMs), most notably chloroform, were produced in the chlorination disinfection process. THM production was proportional to the applied dose. When applying 18 mg/L of chlorine (Table ES-2), the resulting chloroform production was 4.0 μg/L. The chlorine dioxide disinfection process did not produce THMs. The EPA National Primary Drinking Water Regulation's maximum contaminant level (MCL) for Total THMs in drinking water is 80 μg/L.

Both the chlorination and chlorine dioxide disinfection processes produced haloacetic acids (HAAs). When applying 18 mg/L of chlorine and 6.6 mg/L of chlorine dioxide (Table ES-2), the resulting total HAA production was 13μg/L and 36 μg/L for chlorination and chlorine dioxide, respectively. The MCL for HAAs in drinking water standards is 60 μg/L.

It is important to note that the MCLs referenced are for drinking water; unless the effluent of a disinfected, wet weather flow discharge is consumed, these limits should be assumed to be conservative. Also, dilution of disinfection residuals and DBPs by a receiving water should be considered when comparing the values from this demonstration to the MCLs. An approach for addressing receiving water dilution was applied before performing the screening-level risk assessment on the disinfection residuals and DBPs.

For the chlorination disinfection process effluent, all events with total residual chlorine (TRC) less than the detection limit after dechlorination had little or no aquatic toxicity. Events with measurable TRC after dechlorination had significant aquatic toxicity. The data indicate that aquatic toxicity problems can be avoided with proper dechlorination prior to discharge.

For the chlorine dioxide process, aquatic toxicity corresponded to residual chlorine dioxide and chlorite concentrations. Seven of the nine events had 100% mortality for the aquatic test organisms; the mortality rate decreased in the last two events, both of which had non-detected concentrations of residual chlorine dioxide and low chlorite concentrations. The correlation between residual chlorine dioxide and chlorite, and the Microtox EC_{50} is strong ($R^2 > 0.83$) and demonstrates that aquatic toxicity increases with increases in chlorine dioxide and chlorite. Aquatic toxicity decreased in events when ferrous sulfate was used to reduce the chlorine dioxide and chlorite residuals, which indicates that toxicity problems can be avoided with proper control prior to discharge. It should be noted that for the small-scale pilot project, chlorine dioxide was produced by acidifying sodium chlorite, which can contain residual chlorite in the product. Certain methods of chlorine dioxide generation applicable to full-scale facilities do not produce chlorite as a residual to generation; therefore, by using these methods of generation; aquatic toxicity could be avoided without the addition of reducing agents.

For the ozone and UV pilots, no significant aquatic toxicity was measured.

ES.5 Applications for Full-Scale Facilities

Adjusting Data for Site-Specific Conditions

The data collected at the METRO wastewater treatment plant (WWTP) during the demonstration needed to be related to site-specific characteristics of CSO discharges. For example, for this demonstration project, the combined sewage that occurred at the METRO WWTP during wet weather had fairly high COD levels with relatively little variability, compared to CSO in the upper collection system that is known to have diluted COD levels and highly variable pollutant loadings. In addition, the relationship of the CSO discharges to the receiving water was assessed so that the impact of bacteria and DBPs on the environment was more specifically determined.

The disinfection demonstration results were based on the following conditions:

- The subject wet weather flow was collected at the METRO WWTP, which generally has higher average concentrations of pollutants, with the possible exception of grit, than CSOs do.
- Concentrations of disinfection residuals and DBPs did not account for receiving water dilution.
- Disinfection doses were based on meeting the draft EPA guidance values (U.S. EPA 2002) in the end-of-pipe discharge of:
  — fecal coliform density of 4,274 CFU/100 mL
  — *E. coli* density of 2,507 CFU/100 mL
  — *Enterococcus* spp. density of 656 CFU/100 mL.

The data collected during the disinfection demonstration were adjusted to represent site-specific conditions in Onondaga County and utilized to aid the decision-making framework with the local stakeholders.Adjustments were made to the disinfectant doses to meet site-specific bacteria criteria. Once site-specific disinfectant doses were determined, corresponding end-of-pipe and receiving water concentrations of disinfection residuals and DBPs were calculated. An assessment of toxicity was then based on the site-specific concentrations of disinfection residuals and DBPs.

Adjustments were made to the disinfection doses to reflect water quality conditions at the CSO locations as opposed to the water quality of the Metropolitan Syracuse treatment plant during wet weather. Then the dose was adjusted for the dilution attributed to receiving-water conditions from the creeks and Onondaga Lake to achieve a bathing beach criteria of 200 FC/100 ml. The above factors translated to an end-of-pipe concentration of 2,000 FC/100 ml. This target formed the basis of the Onondaga County consent order and is slightly more stringent than the 4,274 FC/100 ml from the BEACHES ACT. Nevertheless, the overriding influence of water quality difference at the CSO locations resulted in decreases in doses from the demonstrating testing values shown in Table ES-2 to the site-specific values shown in Table ES-3. The Onondaga case may serve as a good example of typical state quidance as opposed to the new criteria in the BEACHES ACT based on higher risk.

Once the new targeted disinfectant doses were established for site-specific conditions, new end-of-pipe concentrations of disinfection residuals and DBPs were established, based on the site-specific doses. Table ES-3 presents the site-specific doses and resulting effluent residuals and DBPs. Only residuals and DBPs that increased in concentration during the disinfection processes are presented in Table ES-3.

Table ES-3. Site-Specific Doses and Resulting Residuals and DBPs in the Effluent.

Parameter/Technology	Chlorine	Chlorine Dioxide	Ozonation	UV
Dose[1] (mg/L)	9	4	6	60[2]
Residual w/o Reducing Agent (mg/L)	2.1	0.04	ND[3]	N/A[4]
Chlorate (mg/L)	<1	2.2	NM[5]	NM
Chlorite (mg/L)	ND	14	NM	NM
Chloroform (µg/L)	2.3	ND	NM	NM
Monochloroacetic acid (µg/L)	ND	4.1	NM	NM
Dichloroacetic acid (µg/L)	10	28	NM	NM
Trichloroacetic acid (µg/L)	8.2	8.3	NM	NM

[1] Dose required to achieve 2,000 CFU/100 mL fecal coliforms in effluent
[2] Dose units mWsec/cm^2
[3] Measured, but not detected
[4] Not applicable
[5] Not measured

Effluent concentrations of disinfection residuals and DBPs associated with each disinfection technology were then adjusted to predict environmental concentrations using site-specific receiving water characteristics. This extrapolation was based on receiving water dilution factors. Two receiving waters were considered, Onondaga Creek and Onondaga Lake.

Most of the county's CSOs discharge directly into Onondaga Creek, which is host to many species of fish and invertebrate. A conservative dilution approach was used that assumed that the one-year return CSO event would occur during the minimum average seven-consecutive-day low flow with a 10-year return. This approach resulted in no dilution being offered by Onondaga Creek; in other words, the majority of the flow in the creek during this occurrence is the CSO. However, based on a bacteria model, only approximately 50% of the CSO volume would require disinfection to meet the bacteria requirements as set forth in the consent order. Although there is no dilution offered by the background creek flow, the non-disinfected CSO volume has the effect of diluting residuals and DBPs in the disinfected flows. The resulting dilution factor for Onondaga Creek was calculated to be 0.42.

The second receiving water that was considered was Onondaga Lake, which is designated a recreational water and is also host to many species of fish and invertebrates. The extreme southern end of the lake is considered a mixing zone, which offers dilution in proportion to its volume. The resulting dilution factor in Onondaga Lake was calculated to be 0.008. The calculated environmental concentrations of residuals and DBPs in Onondaga Creek and Onondaga Lake for chlorination and chlorine dioxide are presented in Tables ES-4 and ES-5, respectively.

Table ES-4. Estimated Environmental Concentrations from the Chlorination Process.

Parameter/Technology	Onondaga Creek	Onondaga Lake
Residual without Reducing Agent (mg/L)	0.88	0.017
Chlorate (mg/L)	<0.5	<0.01
Chloroform (µg/L)	1.0	0.018
Dichloroacetic acid (µg/L)	4.2	0.081
Trichloroacetic acid (µg/L)	3.4	0.066

Table ES-5. Estimated Environmental Concentrations from the Chlorine Dioxide Process.

Parameter/Technology	Onondaga Creek	Onondaga Lake
Residual w/o Reducing Agent (mg/L)	0.02	0.0003
Chlorate (mg/L)	0.92	0.018
Chlorite (mg/L)	5.9	0.11
Monochloroacetic acid (µg/L)	1.7	0.032
Dichloroacetic acid (µg/L)	12	0.22
Trichloroacetic acid (µg/L)	3.5	0.066

Onondaga Creek and the extreme southern end of the lake are not used for swimming; therefore, human exposure to disinfection residuals and DBPs in the creek would be incidental. However, aquatic exposure to residuals and DBPs in Onondaga Creek represent an exposure pathway for fish and invertebrates. The Onondaga Lake proper is designated as recreational waters for swimming; therefore, human exposure to disinfection residuals and DBPs was evaluated through a screening-level risk assessment. Fish and invertebrate are also exposed to residuals and DBPs in the lake; however, they are exposed to higher concentrations in the creek. Therefore, the creek concentrations were used for the screening-level risk assessment for aquatic organisms.

Screening-Level Risk Assessment

If there was a significant increase in the concentration of either a residual or DBP in the effluent of a disinfection process, then a screening-level risk assessment was performed on the specific residual or DBP in order to estimate whether the environmental concentration may cause adverse effects in human or aquatic organisms.

In addition to this screening-level risk assessment, aquatic toxicity was measured as part of the disinfection demonstration. Aquatic toxicity was measured in the effluent of each of the disinfection processes.

The screening-level risk assessment included the following steps, which are consistent with the general risk assessment paradigm established by U.S. EPA (U.S. EPA, 1989):

- Hazard Identification
- Exposure Assessment
- Toxicity Assessment
- Risk Characterization.

Through the hazard identification literature review, likely residuals and DBPs were identified. Once identified, these residuals and DBPs were further researched through the literature to obtain toxicity information relevant for assessing risks to human health and the aquatic ecosystem. These findings were used to establish risk-based concentrations (RBC) as toxicity reference values. No new toxicity reference values were derived during this screening-level risk assessment process; only toxicity values found in the literature were used. Assessing exposure included determining what concentrations of these residuals and DBPs were estimated to occur in the environment as well as determining likely exposure pathways. For all the chemicals with sufficient toxicity and exposure data, expected environmental concentrations (EECs) were compared to the RBC values. Chemicals with EECs that exceeded the RBC were characterized as chemicals of potential concern (COPC) because concentrations that exceed risk-based thresholds may pose unacceptable risks to humans or the aquatic ecosystem. Since this was a screening-level assessment, COPC classification means that additional study may be warranted.

In terms of human health risks, there was only sufficient toxicological information about TRC, chlorine dioxide, chlorite, chloroform, and mono- and dichloroacetic acid to assess non-cancer risks and only sufficient toxicological information about chloroform and dichloroacetic acid to assess cancer risks. Both non-cancer and cancer risks were assessed for the estimated environmental concentrations in Onondaga Lake with the concentrations assumed and given in Tables ES-4 and ES-5. Human exposure to residuals and DBPs would occur only in Onondaga Lake and would be unlikely to occur in Onandaga Creek because swimming is permitted in the lake and not in the creek. Non-cancer and cancer risks associated with oral ingestion and inhalation of these chemicals were well below levels of concern, based on exposure due to swimming. However, non-cancer risks due to dermal contact when swimming were elevated for TRC resulting from the chlorination process and for chlorite and dichloroacetic acid resulting from the chlorine dioxide disinfection process. Additionally, dermal contact cancer risks for chloroform and dichloroacetic acid resulting from the chlorine dioxide disinfection process were elevated but below levels of concern. This suggests that additional studies are not needed for cancer risks associated with chloroform and dichloroacetic acid but may be warranted for non-cancer risks associated with TRC, chlorite, and dichloroacetic acid in order to develop more refined estimates of both dermal exposure and toxicity. It is likely that the elevated risk for these chemicals was the result of the conservative assumptions due the uncertainties associated with dermal exposure mechanics. As more is learned about the dermal exposures that are likely at the site, the exposure assumptions may be refined. It is possible that these chemicals will no longer be of concern. It is also important to note that none of these chemical concentrations exceeded the EPA's MCL for drinking water.

In terms of aquatic health risks, residual chlorine from both chlorination and chlorine dioxide disinfection were designated as COPCs. In addition to residual chlorine, chlorite was also designated a COPC for the chlorine dioxide disinfection process. Residual chlorine may be controlled successfully by using a proven reducing agent such as sodium bisulfite.. Chlorite and residual chlorine from chlorine dioxide disinfection can be diminished by using a generation system that limits the production of these chemicals. However, these chemicals can also be controlled successfully by using a reducing agent such sodium bisulfite and ferrous sulfate.

Conceptual Costs

Conceptual costs were developed for chlorination/dechlorination, chlorine dioxide, ozonation, and UV for the doses required to achieve the Onondaga County end-of-pipe bacterial criterion of 2,000 CFU/100 mL fecal coliform. For each technology, a cost was developed for 1, 10, and 100 million gallons per day (MGD) peak flow facility. Chlorination/dechlorination and chlorine dioxide were cost competitive for the site up to flows of 100 MGD. UV appeared to be cost competitive at 1 MGD and 10 MGD. At a flow rate of 100 MGD, UV was not cost competitive as compared to chlorination/dechlorination and chlorine dioxide. Ozone was slightly more costly than the other technologies at the 1 MGD flow rate not cost competitive at the 10 MGD and 100 MGD flow rate. The high cost of ozonation can be attributed to the complexity of injecting ozone into the wet weather flow and the cost of producing ozone. Even the most efficient ozone generators require proportionally much greater electrical costs than the other technologies.

ES.6 Communication

During this project, an important component of the decision-making framework was implementing a communication plan for stakeholder involvement. Local stakeholders were involved throughout the project by means of monthly progress reports and workshops. Not included in the decision-making framework but equally important was public outreach. The public was involved through a public workshop.

The stakeholders were involved in the beginning to ensure that all information relevant to the assessment was identified, after the disinfection demonstration to be appraised of the results, after data analysis to be updated with the benefit/risk/feasibility assessment results, and at the end to help make decisions about technology selection.

In addition to the stakeholder workshop, the general public was invited to participate in a public workshop where local concerns and perceptions were voiced regarding the tradeoffs between microbial risks to people versus chemical risks to people and to the aquatic ecosystem. Concerns brought forth by the public included: location of wet weather flow treatment facilities in residential areas, odors from the facility, safe handling of disinfectants at the facility, and the discharge of disinfected CSOs with DBPs.

Selecting local stakeholders to participate in the decision-making process was an effective approach in soliciting local values and concerns in shaping the decisions regarding the selection of appropriate disinfection technologies. Even though the local stakeholders represented a broad range of views, none of the stakeholders were entirely opposed to the idea of disinfecting wet weather flows. The public workshop presented an opportunity for communicating preliminary findings of this project to the public. Many good questions were asked along with statements of concerns and issues. Some attendees expressed opposition to the idea of treating CSOs and discharging the treated CSOs into the creek and thus presented the opportunity to educate the public about wet weather flow and disinfection.

Although the risks associated with wet weather flow disinfection were found to be low, select public opposition was high based on reactions from the workshop, . A communication plan was prepared at the onset of the project that included "Stakeholder Relations," but did not include a program to communicate with opposition groups within the general public. Such a program would need to include both supporters and critics from the public as well as stakeholders throughout the decision-making process in order to help gain trust along the way..

ES.7 Organization of Report

Chapter 1.0, Introduction introduces the rationale for wet weather flow disinfection and defines different types of wet weather flow including combined sewer overflow, sanitary sewer overflow and stormwater. It also outlines the objectives and scope of the project.

Chapter 2.0, Decision-Making Framework presents the approach used in this project to select appropriate disinfection technologies for wet weather flow. This decision-making framework approach could be used at the national, state, and local level to guide wet weather flow disinfection control policies and technology selection.

Chapter 3.0, Literature Review presents technical descriptions of disinfection technologies, including chlorination, chlorine dioxide, ozonation, UV, peracetic acid, E-beam, bromine, and mixed oxidants. The literature review also provides citations that document the benefits of disinfection throughout history as well as the more recent recognition of the risks associated with disinfection residuals and by-products. A hazard identification was performed as part of the risk assessment for the disinfection residuals and by-products identified in the literature review.

Chapter 4.0, Disinfection Demonstration presents the findings from the disinfection demonstration, including the required dose to achieve U.S. EPA's *E. coli* bacterial criterion as well as the concentrations and associated toxicity of disinfection residuals and by-products produced during the disinfection demonstration.

Chapter 5.0, Applications for Full-Scale Facilities provides an example of how to use the data derived from *Chapter 4.0* to develop design parameters and operational settings, as well as issues pertaining to assessing risks and developing conceptual costs.

Chapter 6.0, Communication presents the value of a communication program and provides recommendations for developing and implementing a communication plan for projects related to disinfection.

Chapter 7.0, Conclusions summarizes the findings from the three phases of this project.

Chapter 8.0, Hazard Identification provides a literature review pertaining to the human and aquatic organism effects of disinfection residuals and by-products.

CHAPTER 1.0

INTRODUCTION

1.1 Rationale for Wet Weather Flow Disinfection

Urban wet weather pollution manifests itself in three different ways: combined sewer overflows (CSO), sanitary sewer overflows (SSO), and stormwater (SW) runoff. To minimize the impact associated with these occurrences, disinfection of wet weather flows is generally practiced to control the discharge of pathogens and other microorganisms to receiving waters. Disinfection typically relies on the addition of powerful oxidants such as chlorine in the form of gas or sodium hypochlorite (NaOCl), chlorine dioxide (ClO_2), or the use of ozonation or irradiation of the wastewater with ultraviolet light (UV).

Wet weather flows originate from a range of sources including diffuse overland flows, storm flows from industrial and municipal storm sewers, CSOs, and SSOs. U.S. EPA defines these wet weather flows using the descriptions offered in the following subsections.

1.1.1 Stormwater

Urban SW runoff, which is often collected by storm drains and transported to receiving waters, contains pollutants that are accumulated in rainwater or snowmelt flows across roads and the surface of the earth. Such pollutants include oil and grease, chemicals, nutrients, heavy metals, bacteria, viruses and oxygen-demanding compounds (U.S. EPA, 2000a).

1.1.2 Combined Sewer Overflows

Combined sewer systems collect both SW runoff and sanitary sewage in the same pipe. During dry weather these combined sewer systems transport wastewater directly to the sewage treatment plant. In periods of rainfall or snowmelt, however, the wastewater volume in a combined sewer system can exceed the capacity of the sewer system or treatment plant. In this instance, combined sewer systems are designed to overflow occasionally during wet weather periods and discharge excess wastewater directly to nearby streams, rivers, lakes, or estuaries. CSOs contain SW and untreated human and industrial waste, toxic materials, and debris (U.S. EPA, 2000b).

1.1.3 Sanitary Sewer Overflows

SSOs are designed to collect and convey only sanitary sewage from their points of origin to wastewater treatment plants (WWTPs). SSOs are discharges of raw sewage from these systems. SSOs can release untreated sewage into basements, city streets, public and privates lands, and nearby streams, rivers, lakes, or estuaries. SSOs occasionally occur in almost every sewer system even though systems are intended to collect and contain all the sewage that flows into them. Problems that can cause SSOs include:

- infiltration and inflow (I&I);

- undersized systems;
- pipe failures; and
- equipment failures (U.S. EPA, 2000c).

1.1.4 Wet Weather Flow Disinfection

Most wet weather flows can be characterized as having high organic content and containing such pollutants as oil and grease, oxygen-demanding compounds, chemicals, nutrients, heavy metals, bacteria, and viruses. A significant effect of untreated wet weather flows on receiving water quality is contamination with human disease-causing organisms, referred to herein as pathogenic or enteric microorganisms.

As a means to reduce pathogenic contamination, the U.S. EPA in its May 2002 draft, *Implementation Guidance for Ambient Water Quality Criteria for Bacteria*, cites specific allowable water quality standards for indicator species. A summary of these standards is presented in the literature review (Chapter 3.0).

From the perspective of human health protection exposure to pathogens and the associated health concerns, the disinfection process is arguably the most important stage of wet weather flow treatment. The benefits of disinfecting wet weather flow, in terms of reducing pathogen concentrations, have been demonstrated numerous times over the past 30 years (U.S. EPA, 1975; U.S. EPA, 1979a, U.S. EPA, 1979b; Camp Dresser & McKee and Moffa & Associates, 1997). However, disinfectants added to wet weather flows can result in residuals and react with organic matter in the water to form substances known as disinfection by-products (DBPs). The types and quantity of residuals and DBPs is a function of the disinfectant type and dose, the length of contact time, and the water quality characteristics inherent to the wet weather flow. Consequently, optimization is important to balance the maximum protection of public health with minimal disinfectant dose.

In most cases, extensive research regarding disinfection residuals and DBPs is related to the use of chlorine as a disinfectant for drinking water. Chlorination has proven to be an effective and inexpensive means of disinfection and consequently is the most widely used. However, the use of chlorine has been shown to result in the formation of trihalomethanes (THMs), which may be carcinogenic to humans. As a result, regulations have been enacted to limit the quantity of chlorine used to disinfect drinking water. Conversely, due to limited information pertaining to the risks associated with disinfecting wet weather flows, there is very little in the way of regulations. Risks associated with disinfecting wet weather flows include potential contamination of downstream drinking water supplies with disinfection residuals and DBPs, as well as human contact with disinfection residuals and DBPs through recreational exposure. Aquatic organism exposure to disinfection residuals and DBPs in the receiving waters is also a risk.

The necessity to provide public health protection from wet weather pollution has created a need to examine the benefits and potential risks resulting from disinfection treatment of wet weather flows. This includes the evaluation of the traditional methods of chlorination, as well as alternative methods such as chlorine dioxide, ozonation, and UV irradiation. CSO, SSO, and SW control programs are being implemented across the United States to decrease the environmental impacts associated with wet weather pollution. As a result, it is important to develop an informative and useful database on the benefits and risks associated with disinfecting wet weather flows. A better understanding of the relative effectiveness of disinfection technologies, as well as their impacts on human health and the aquatic environment, will help define effective control programs for the protection of human health and, at the same time, avoid adverse impacts on the environment.

1.2 Objectives and Scope of Project

The primary objective of this project was to identify and communicate the benefits and risks associated with disinfecting wet weather flows with chlorine and viable alternative disinfectants, including chlorine dioxide, ozonation, and UV. The benefits of disinfection include the reduction of pathogens discharged to receiving waters that may be used for recreation or municipal drinking water supplies. The risks of disinfection include an increase in disinfection residuals and DBPs to these waters.

The three major phases of this project were:

1. *Literature Review* Gather and summarize existing information on the effectiveness (i.e., benefits) of wet weather flow disinfection technologies and the associated disinfection residuals and DBPs and their potential risks to aquatic and human life.
2. *Disinfection Demonstration* Perform a disinfection demonstration to evaluate the effectiveness (i.e., benefits) of available disinfection technologies and measure associated disinfection residuals and DBPs.
3. *Communication* Communicate the findings of the literature review and disinfection demonstration to stakeholders and the public.

The three phases of this project were implemented in collaboration with Onondaga County Department of Water Environment Protection and local stakeholders. A decision-making approach was developed and implemented for the purpose of evaluating the benefits and risks associated with disinfection technologies used for wet weather flows using site-specific information from the Onondaga County case study.

This report includes selected findings from the literature review, disinfection demonstration, and communication phases of this project.

CHAPTER 2.0

DECISION-MAKING FRAMEWORK

2.1 Risk Management

Owing to the highly variable nature of wet weather flow discharges and the consequent public health and aquatic impacts of both the flows themselves and the use of disinfectants (safe handling and DBPs), there is a need to employ risk management to accomplish desired objectives.

Risk management is a process of making informed decisions about how to apply technology and resources to mitigate risks. A successful risk management process is one that:

- Facilitates the flow of useful information to decision and policy makers;
- Clearly defines and substantiates risk management goals;
- Successfully implements engineering or other processes to achieve these goals; and
- Successfully communicates the process to those affected by the risk management decisions (i.e., stakeholders).

The decision-making framework shown in Figure 2-1 is designed to achieve these goals.

2.2 Decision-Making Framework

The decision-making framework consists of three interacting components, namely:

1. Information Inputs (including site-specific inputs);
2. Benefit/Risk/Feasibility Assessment; and
3. Technology Selection.

These components are depicted in the left, center, and right blocks, respectively, of Figure 2-1. The three components form decision-feedback loops, which can be repeated as necessary to satisfy risk management needs.

2.2.1 Information Inputs

Information inputs to the risk management of wet weather flow disinfection include exposure-related and hazard-related inputs, shown in the upper left and lower left block of Figure 2-1, respectively. The exposure-related inputs include the chemical and hydrodynamic characteristics of the wet weather flows and receiving waters, as well as the potential exposure pathways by which human and aquatic organisms may come into contact with pathogens or disinfection residuals and DBPs. By their nature, the inputs are site-specific. Therefore site-specific information from the Onondaga County case study was used while implementing the decision-making framework for this project. However, there are many general findings that can be useful for other municipalities and wet weather flow types. The hazard-related inputs include the identification of pathogenic organisms and their related diseases and toxicity of potential disinfection residuals and DBPs. These inputs form the basis of a conceptual model that represents both the benefits and risks associated with disinfecting or not disinfecting wet weather flows and exposure pathways relevant to a given wet weather flow and receiving water.

Initial site-specific information can be obtained for other municipalities from readily available literature and from consultation with stakeholders who may have access to data that is not in the public domain.

2.2.2 Benefit/Risk/Feasibility Assessment

The initial phase of the benefit/risk/feasibility assessment (left side of the middle block of Figure 2-1) involves assessing the adequacy of information on efficacy, hazards, and benefit/risk benchmarks. Benefit/risk benchmarks refer to quantitative indices of acceptable levels of benefit and risk. For example, these may be national, state, or local standards for water quality or public health. Where no such standards exist, site-specific benchmarks should be developed. Examples of these are maximum levels of pathogens leaving the facility or entering the receiving water, risk-based maximum concentrations of disinfectant residuals, or DBPs leaving the facility or entering the receiving water.

Data gaps identified in the initial phase of the benefit/risk/feasibility assessment (indicated with a "No" decision on the left side of the middle block of Figure 2-1) become goals for further data collection, which may take the form of a disinfection demonstration. Such data may provide definitive information on toxicity and benefits resulting in a "Yes" answer to these questions. If sufficient information cannot be obtained, then stakeholder input is required to gain consensus for the benefit/risk/feasibility assessment process to move forward. This process may be repeated as many times as needed to complete the benefit/risk assessment to the satisfaction of policy makers and stakeholders.

Once critical data gaps are filled and benchmarks are established, a benefit/risk assessment can be conducted. If benefit and risk goals are projected to be achieved for a given technology, the engineering feasibility and costs can be determined.

2.2.3 Technology or Policy Selection

Based on the outcomes of the benefit/risk/feasibility assessment, a technology or policy may be selected for implementation as a solution. Other technologies/policies may be assessed using the same framework (right block of Figure 2-1).

2.2.4 Stakeholder Involvement

An important component of the framework is stakeholder involvement and communication. In the case of this decision-making framework, stakeholders include informed citizens and regulatory personnel. Stakeholder involvement is shown at four places in the decision process, although this may vary depending on stakeholder interest.

Figure 2-1 shows stakeholder involvement being sought at the following four stages:

1. *Site-Specific and General Information Input* to ensure that all information relevant to the assessment is identified;
2. *Benefit/Risk/Feasibility Assessment-Data Collection* to ensure that assumptions regarding the wet weather flow discharges and receiving water are valid and to plan for collecting data through a disinfection demonstration;
3. *Benefit/Risk/Feasibility Assessment-Review of Data* to communicate the findings of the disinfection demonstration and begin *Benefit/Risk Assessment;* and
4. *Technology Selection* to inform stakeholders of risk management decisions and exclude disinfection technologies that do not meet criteria.

This stakeholder involvement plan is common for typical wastewater treatment programs today. The stakeholders are involved in the beginning to ensure that all information relevant to the assessment is identified, in the middle so that they can be apprised of the results of data collection, and at the end so that they may learn and help make the final decisions. The public may also be included in the decision-making process through a public workshop. This is where local values and perceptions of benefits and risk are voiced regarding the tradeoffs between microbial risks to people versus chemical risks to people and to the aquatic community.

This project was completed in collaboration with Onondaga County Department of Water Environment Protection and local stakeholders. Local stakeholders, including citizens and regulatory personnel, were involved throughout the project as noted above. The local stakeholders were also included in the decision-making process through a public workshop where local values and perceptions regarding wet weather flow disinfection were presented. The local stakeholders were:

Richard Elander, P.E.
Onondaga County Department of Water Environment Protection

Stephen Martin
Onondaga County Department of Water Environment Protection

Thomas Friedrich
Trout Unlimited

Michael Pacelli
Concerned citizen representative

Russell Nemachek, P.E.
Onondaga County Department of Health

Elizabeth Colye, P.E.
New York State Department of Environmental Conservation, Region 7

John Ferrante, Ph.D.
Central New York Regional Planning & Development Board

James Hassett, Ph.D.
State University of New York College of Environmental Science and Forestry

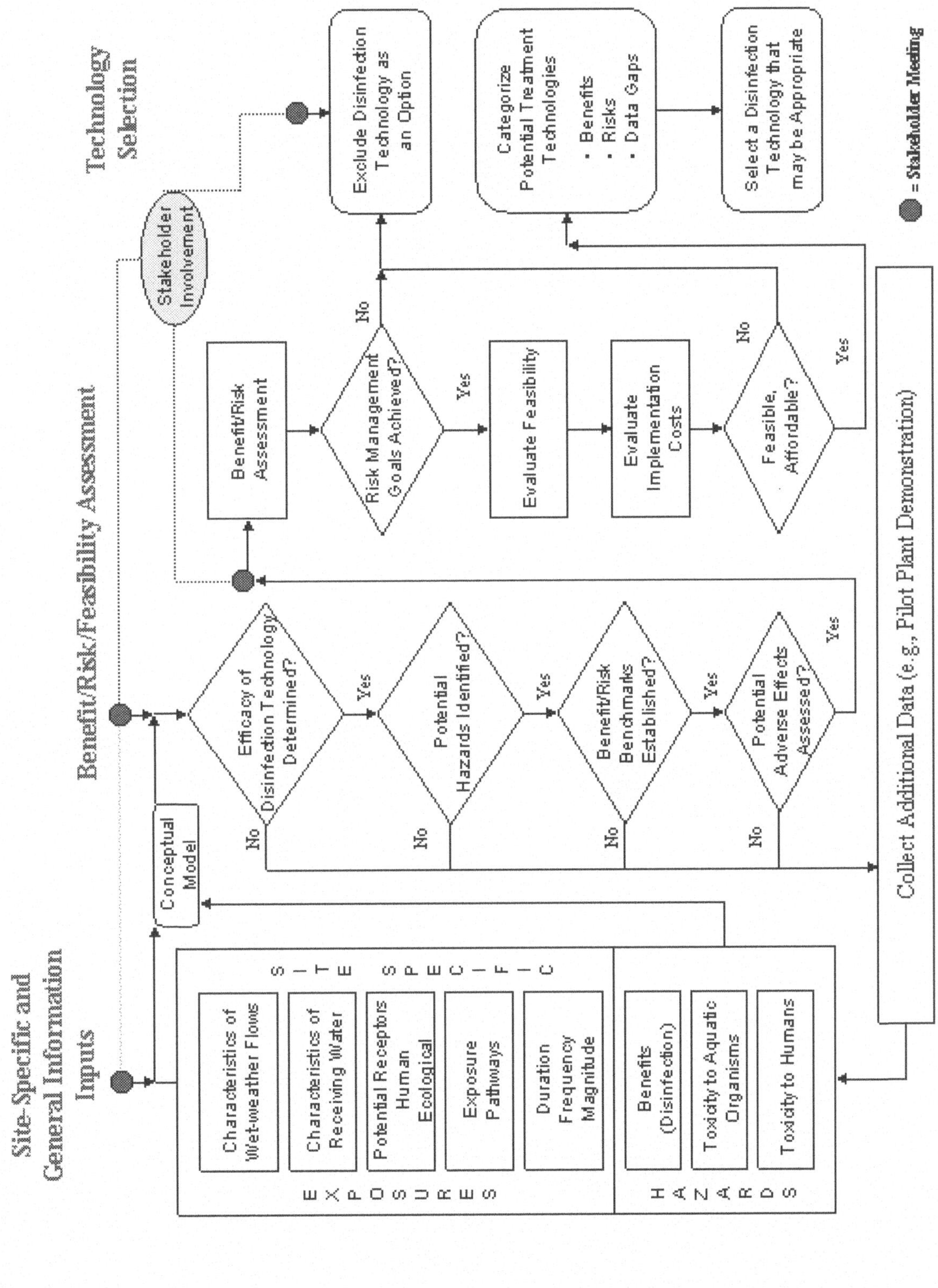

Figure 2-1. Decision-Making Framework.

CHAPTER 3.0

LITERATURE REVIEW

3.1 Wet Weather Flow Disinfection Technology

The U.S. EPA has published many fact sheets on the topic of CSOs, which also may be applicable for SW and SSO in terms of disinfection. *Combined Sewer Overflow Technology Fact Sheets* can be downloaded from the U.S. EPA Website (www.epa.gov). Two such CSO Technolgy Fact Sheets are *Chlorine Disinfection* (U.S. EPA, 1999b) and *Alternative Disinfection Methods* (U.S. EPA, 1999a).

Chlorine has long been the most common disinfectant for drinking water, wastewater, and wet weather flows. More recently, alternatives to chlorine have gained recognition. The alternatives identified in the U.S. EPA *Combined Sewer Overflow Technology Fact Sheet, Alternative Disinfection Methods* are:

- Chlorine dioxide;
- Ozonation;
- UV radiation;
- Peracetic acid; and
- Electron beam irradiation (E-Beam).

Other disinfectants that may also be appropriate for wet weather flows are bromine and mixed oxidants.

The following issues are generally considered when applying these disinfection techniques to wastewater and wet weather flows: 1) types of organisms, 2) number of organisms, 3) nature of suspending liquid, 4) temperature, 5) intensity and nature of physical agent, 6) concentration and type of chemical agent, and 7) contact time (Pelczar, Reid, and Chan, 1986).

Experience has shown that the long contact time required for conventional wastewater treatment is extremely costly for the treatment of wet weather flows due to the very high flow rates. However, disinfection of wet weather flows can be achieved at shorter contact times by providing intense mixing and an increased disinfection dosage to ensure disinfectant contact with the maximum number of microorganisms (Benjes, 1976). Intense mixing can shear particles, thereby increasing the degree of contact with the disinfectant. Mixing can be accomplished by mechanical flash mixers at the point of disinfectant injection and at various points along the length of the contact chambers (White, 1999).

Because of the intermittent and infrequent operation associated with treatment of wet weather flows, reduction of construction costs combined with increased operating costs often result in overall lower costs. In the case of disinfection facilities using chemical oxidants, the high construction costs associated with conventional contact basins (contact times of 15–30 minutes) are difficult to justify. Therefore, consideration has been given to smaller, shorter-detention-time contact basins equipped with larger mixers and two-stage dosing to make better usage of the oxidants and/or higher dosages that produce the same microorganism reductions as conventional basins (Benjes, 1976). Also the use of oxidants (disinfectants) that are capable of disinfecting more rapidly result in a more economical disinfection facility. For example, chlorine dioxide is a more rapid oxidant that requires less contact time than chlorine, and thus reduces the required contact time, basin size, and related capital costs of a disinfection facility (Moffa and Associates, 2000).

Disinfection by oxidants and radiation are often only effective for the microorganisms not embedded in particulate matter or microorganisms on or near the surface of particulate material. Depending on the degree of upstream treatment (solids removal), the disinfection process may not be effective if microorganisms are contained or harbored within larger protective solid particles and fluid droplets. Chemical oxidants are limited in their ability to penetrate, which is a function of chemical demand and harboring of microorganism or occlusion by particles, whereas radiation is limited by transmissivity of the water and microorganism occlusion by particles. These limits are based on the disinfectant concentration/intensity, particle size and make-up, and water chemistry.

3.2 Benefits of Disinfecting Wet Weather Flows

The most important reason for disinfecting water is to inactivate pathogenic organisms, thus limiting waterborne disease. The benefits of disinfecting water, has been well documented since early in the twentieth century and is most well documented in drinking water applications. As more cities adopted water chlorination, U.S. death rates due to typhoid, cholera and hepatitis A also declined dramatically, being virtually eliminated. The filtration and disinfection of drinking water has been responsible for a large part of the 50 percent increase in life expectancy in this century. Worldwide, significant strides in public health and the quality of life are directly linked to the adoption of drinking water chlorination. Documenting advances of the 20th century, Life magazine cited the filtration of drinking water and use of chlorine as "probably the most significant public health advance of the millennium." (Life, 2000) Additionally, the Centers for Disease Control and Prevention and the World Health Organization both recognized water treatment as one of the century's greatest public health achievements (Centers for Disease Control, 1999).

The association between enteric diseases and fecal contamination in drinking water is an obvious one by its very nature. This, in conjunction with a definitive act by the Department of the Treasury to set standards requiring disinfection of drinking water in October of 1914, established a rich history delineating the benefits of disinfecting drinking water.

The benefits of disinfecting wastewater and wet weather flows discharging to receiving waters relate to the receiving waters' use as a drinking water supply or recreational resource. Disinfection of these discharges reduces microbial concentrations of pathogens and related health risks. Nevertheless, outbreaks of disease from drinking water are reported each year

(Water Pollution Control Federation, 1984; Moore et al., 1994) as a result of inadequately disinfected wastewater and wet weather flow pollution.

Although they are not as well documented, the benefits of disinfecting wastewater and wet weather flows that discharge into receiving waters, particularly recreational waters and fisheries, can be inferred in the scientific literature. There has been some debate over whether enteric disease can be transmitted by contact recreation; however, past epidemics and epidemiological studies indicate that recreation is associated with disease transmission (Water Pollution Control Federation, 1984; Moore et al., 1994). For example, a 1974 shigellosis outbreak in Dubuque, Iowa, was related to swimming in the Mississippi River (Rosenberg et al., 1975). The mean level of fecal coliform in the water was 17,500/100 mL, with counts as high as $5x10^6$/100 mL downstream from the WWTP and $4x10^5$/100 mL at the beach.

Research has also been performed in connection with setting recreational water quality criteria for bacterial concentrations. A direct relationship between swimming-associated enteric disease outbreaks and the quality of the bathing water was obtained from a multi-year epidemiological-microbiologic research program conducted in New York City (1973–1975), Lake Pontchartrain, Louisiana (1977–1978), and Boston, Massachusetts (1978). Several microbial indicators were used to define the quality of the water; presence of *Enterococcus* spp. showed the best correlation to the total and highly credible gastrointestinal symptoms. Additionally, the frequency of gastrointestinal symptoms had a high degree of association with the distance from the known sources of municipal wastewater outfalls (Cabelli et al., 1979). The Centers for Disease Control and Prevention has also documented the occurrences of reported waterborne disease outbreaks relating to recreational waters (Levy, 1998; Centers for Disease Control, 2001) when the WWTP influent was inadequately disinfected.

3.3 Risks of Disinfecting Wet Weather Flows—Hazard Identification

The main objective of disinfecting wastewater and wet weather flows is to control the quantity and concentration of pathogens by inactivating or destroying a wide range of bacteria and viruses. This goal has commonly been achieved by using chemicals, primarily chlorine or one of its derivatives (Water Environment Federation, 1996). Although chlorine disinfection has significantly reduced the incidence of waterborne disease in the United States for many years, growing concerns about chlorine's safety and effect on the environment have prompted communities to evaluate other disinfection methods (Haas, 1986). The concerns regarding chlorine disinfection are mainly in two areas: 1) safety in handling and use, and 2) creation of DBPs by the addition of chlorine to the wastewater.

Each year, the National Response Center receives reports on more than 25,000 fires, spills, or explosions involving many types of hazardous chemicals, including chlorine (National Response Center, 2001). In 1998 there were over 100 deaths, nearly 5,000 injuries and,when including small spills, almost 50,000 incidents related to ordinary industrial use of many types of chemicals in the United States (Mannan and O'Connor, 2001). In April of 1996, a train carrying a tank car of chlorine derailed near Alberton, Montana, which resulted in one of the largest chlorine spills in U.S. history, about 100,000 gallons. More than 250 people checked into local hospitals, while an emergency crew evacuated the town of Alberton and closed Interstate 90 for more than two weeks (Missoula County, 2000).

Even though chlorine gas has an excellent safety record, public perception of the impacts of the worst-case scenario has led to major concerns. This public perception, however, is not without cause. The U.S. EPA has published the following hazard summary on chlorine (U.S. EPA, 2000):

> Acute (short-term) exposure to high levels (>30 ppm) of chlorine gas in humans results in chest pain, vomiting, toxic pneumonitis, pulmonary edema, and death. At lower levels, (< 3 ppm), chlorine gas is a potent irritant to the eyes, the upper respiratory tract, and lungs.

As a result of these hazards and the concern that large quantities of chlorine gas could be accidentally or maliciously released from cylinders at treatment plants, many treatment plants have converted from gaseous chlorine to a less volatile form of chlorine, sodium hypochlorite (Orum, 2001).

Beginning in the 1970s and continuing through the 1980s and 1990s, the concerns about DBPs reached significant levels (Symons, 2001). The awareness that THMs were formed when chlorine reacted with organic precursors alerted many people to the quality of their drinking water. This awareness of THMs led to the discovery of other DBPs in the effluent from WWTP. As a result, the U.S. EPA, through the National Pollutant Discharge Elimination System (NPDES), established permit requirements for the allowable total chlorine residual in discharges from WWTP. THMs and other DBPs have been classified by U.S. EPA as "probable human carcinogens" based on evidence provided from animal studies. There is no direct proof that THM in drinking water causes cancer in humans. The primary concern is related to the possible relationship of liver, kidney and bladder cancer with long-term exposure to contaminating THMs (U.S. EPA, 1999c).

Table 3-1 is an EPA list of disinfection residuals and DBPs that may pose health concerns (U.S. EPA, 1999c). The table includes both the disinfectant residuals and the specific DBPs produced by the disinfectants' water treatment. These contaminants of concern are grouped into four distinct categories and include disinfectant residuals, inorganic by-products, organic oxidation by-products, and halogenated organic by-products.

3.3.1 Hazard Identification

A hazard identification was performed for this project for a subset of the chemicals listed in Table 3-1. This hazard identification is presented in Chapter 8.0 and summarized in Table 3-2 for human health and aquatic organism toxicity.

The information summarized in Table 3-2 was determined from the hazard identification process based on potential exposure to disinfection residuals and DPBs following disinfection of wet weather flows. The available toxicity information, as summarized in this table, can be used to perform only a screening-level risk assessment. No new toxicity values were derived during the hazard identification process. Additional research is needed to perform a comprehensive exposure and toxicity assessment on each chemical. For example, in most instances, these chemicals will not occur exclusively. Instead, they are present in wet weather flows as complex mixtures depending on the treatment and characteristics of both the source water and receiving water (WHO, 2000). The toxicology of many of these chemicals has not been studied in great detail, and even less is known about their toxicity when they occur as mixtures. Nevertheless, there are some empirical data on mixtures, and risk assessment guidance for mixtures is currently being developed by both the U.S. EPA and the Agency for Toxic Substances and Disease Registry (U.S. EPA, 2000d; ATSDR, 2001).

Table 3-1. Disinfection Residuals and DBPs (U.S. EPA, 1999c).

DISINFECTANT RESIDUALS

- Free Chlorine
  - Hypochlorous Acid
  - Hypochlorite Ion
- Chloramines
  - Monochloramine
- Chlorine Dioxide

INORGANIC BY-PRODUCTS

- Chlorate Ion
- Chlorite Ion
- Bromate Ion
- Iodate Ion
- Hydrogen Peroxide

ORGANIC OXIDATION BY-PRODUCTS

- Aldehydes
  - Formaldehyde
  - Acetaldehyde
  - Glyoxal
  - Hexanal
  - Heptanal
- Carboxylic Acids
  - Hexanoic Acid
  - Heptanoic Acid
  - Oxalic Acid
- Assimilable Organic Carbon

HALOGENATED ORGANIC BY-PRODUCTS

- Trihalomethanes
  - Chloroform
  - Bromodichloromethane
  - Dibromochloromethane
  - Bromoform
- Haloacetic Acids
  - Monochloroacetic Acid
  - Dichloroacetic Acid
  - Trichloroacetic Acid
  - Monobromoacetic Acid
  - Dibromoacetic Acid
- Haloacetonitriles
  - Dichloroacetronitrile
  - Bromochloroacetonitrile
  - Dibromoacetonitrile
  - Trichloroacetonitrile
- Haloketones
  - 1, 1 -Dichloropropanone
  - 1, 1, 1 -Trichloropropanone
- Chlorophenols
  - 2-Chlorophenol
  - 2,4-Dichlorophenol
  - 2,4,6-Trichlorophenol
- Chloropicrin
- Chloral Hydrate
- Cyanogen Chloride
- N-Organochloramines
- 3-Chloro-4-(dichloromethyl)-5-hydroxy-2(5H)-furanone

Table 3-2. Disinfection Residual and DBP Hazard Identification Summary.

Chemical	CAS NO	Disinfection Process			Potential Effects	
		Chlorination	Chlorine Dioxide	Ozone	Human Health	Aquatic Ecological
Residuals						
Hypochlorous acid	7790-92-3	Y			Y	Y
Hypochlorite ion	14380-61-1	Y			Y	Y
Monochloramine	10599-90-3	Y			Y	Y
Dichloramine	3400-09-7	Y				Y
Trichloramine	10025-85-1	Y				Y
Chlorine dioxide	10049-04-4		Y		Y	Y
Inorganic By-Products						
Chlorate ion (chlorate hydrate)	14866-68-3		Y	Y		Y
Chlorite ion	7758-19-2	Y	Y		Y	Y
Bromate ion	15541-45-4	Y		Y	Y	Y
Iodate ion						Y
Potassium, salt	7758-05-6			Y		Y
Sodium salt	7681-55-2			Y		Y
Hydrogen peroxide	7722-84-1			Y		Y
Ammonia	7664-41-7					Y
Aldehydes						
Formaldehyde	50-00-0	Y		Y	Y	Y
Acetaldehyde	75-07-0	Y		Y		Y
Glyoxal	107-22-2	Y		Y		Y
Meythylglyoxal	78-98-8	Y		Y		Y
Benzaldehyde	100-52-7	Y		Y		Y
Propanal	123-38-6	Y		Y		Y
Aldehydes						
Hexanoic acid (Caproic acid)	124-62-1	Y		Y		Y
Heptanoic acid	111-14-8	Y		Y		Y
Oxalic acid (metabolite of DCA)	144-62-7	Y		Y		Y
Hexadecanoic acid (Palmitic acid)	57-10-3	Y		Y		Y
Butanedioic acid (Succinic acid)	110-15-6	Y		Y		Y
Benzoic acid	65-85-0	Y		Y		Y
Benzene dicarboxylic acid (3 isomers)	29010-86-4	Y		Y		Y
Benzene tricarboxylic acid (3 isomers)	569-51-7	Y		Y		Y
Methylfuran carboxylic acid		Y		Y		Y
Trihalomethanes						
Chloroform	67-66-3	Y			Y	Y
Bromodichloromethane	75-27-4	Y			Y	Y
Dibromochloromethane	124-48-1	Y			Y	Y
Bromoform	75-25-2	Y			Y	Y
Trihalomethanes						
Monochloroacteic acid (metabolite of DCA)	79-11-8	Y				Y
Dichloroacetic acid (DCA)	79-43-6	Y			Y	Y
Trichloroacetic acid	76-03-9	Y				Y
Monobromoacetic acid	79-08-3			Y	Y	Y
Dibromoacetic acid	631-64-1			Y	Y	Y
Haloacetonitriles						
Dichloroacetonitrile	3018-12-0	Y				Y
Bromochloroacetonitrile	83463-62-1	Y				Y
Dibromoacetonitrile	3252-43-5	Y				Y
Trichloroacetonitrile	545-06-2	Y			Y	Y
Haloketones						
1,1-Dichloropropanone	513-88-2			Y		Y
1,1,1-Trichloropropanone	918-00-3			Y		Y
Chlorophenols						
2-Chlorophenol	95-57-8	Y				Y
2,4-Dichlorophenol	120-83-2	Y			Y	Y
2,4,6-Trichlorophenol	88-06-2	Y			Y	Y
Other						
Sodium Bisulfite (Chlorine quenching agent)	7631-90-5	Y			Y	Y
Chloropicrin	76-06-2	Y				Y
Chloral hydrate	302-17-0	Y				Y
Cyanogen Chloride	506-77-4	Y				Y
3-Chloro-4-(dichloromethyl)5-hyrdoxy-2(5H)-furanone (MX)	77439-76-0	Y				Y

Note: "Y" indicates presence or potential presence, but does not indicate magnitude or potential risk.

3.4 Bacterial Standards

Indicator bacterial organisms are used to predict and/or minimize the potential risk from pathogenic microbes. Indicator organisms are useful in that they circumvent the need to assay for each and every pathogen. Ideally, indicators are rapidly detected, easily enumerated, and have survival characteristics that are similar to those of the pathogen of concern. Fecal coliforms have been used extensively for many years as indicators of fecal contamination. However, the coliforms' ecology, prevalence, and resistance to stress differ from many of the pathogenic organisms for which they are proxy. These differences limit the applicability of fecal coliform as an indicator organism. Therefore, numerous alternative microbes have been suggested, including *Escherichia coli*, *Enterococcus* spp., *Clostridium perfringens*, coliphages, and bifidobacteriaphages (American Academy of Microbiology, 2001).

The U.S. EPA and a number of states are currently modifying bacteria standards for the protection of public health. Indicator bacterial organisms of potential pathogen contamination include fecal coliform, *E. coli* and *Enterococcus* spp. Current standards for most regulatory authorities use fecal coliform. The U.S. EPA in its May 2002 draft, *Implementation Guidance for Ambient Water Quality Criteria for Bacteria*, recommends the use of the latter two indicators to assess inland and marine water quality, respectively.

The draft guidance defines bacterial indicator concentrations for different illness rates. In the past, a rate of 8 illnesses per 1,000 people was used for freshwater and 19 per 1,000 people for marine waters. In the guidance, the U.S. EPA will allow states to adopt a freshwater standard rate of up to 14 illnesses per 1,000 people. The criteria are based upon beach and lake epidemiological investigations that are referenced by the U.S. EPA in *Implementation Guidance for Ambient Water Quality Criteria for Bacteria.* Geometric mean values, maximum values, and illness rates are displayed in Table 3-3.

Table 3-3. Draft U.S. EPA Indicator Bacterial Organisms Criteria.

Indicator Organism	Geometric Mean	Maximum Value	Illness Rate
Fecal coliform	840 CFU*/100 mL	4,274 CFU/100 mL	14 per 1,000
E. coli	548 CFU/100 mL	2,507 CFU/100 mL	14 per 1,000
Enterococcus spp.	144 CFU/100 mL	656 CFU/100 mL	19 per 1,000

* colony-forming unit

Note: The maximum value for fecal coliform is not specified by the U.S. EPA. It was calculated using data collected from this disinfection demonstration. A relation between *E. coli* and fecal coliform was developed using the ratio of these organisms measured in the disinfected effluent. This relationship was then used to calculate the fecal coliform concentration that corresponded to the *E. coli* criterion of 2,507 CFU/100 mL, which was 4,274 CFU/100 mL.

Both genetic and immunological tools are available and have been used to address waterborne pathogenic microorganisms. Many researchers are developing "gene probes" to detect pathogenic organisms' genes. Gene probes are being used that are highly specific and capable of detecting genetic sequences of DNA and RNA common to pathogens such as *salmonella* and *legionella*. Genotyping, or molecular characterization, is a powerful tool for source identification of microbial contamination. Molecular fingerprints, host-specific genes, and chemical constituents provide direct evidence of source origin, while general indicators provide virtually no information about sources. Although new techniques show promise, their application must be specifically defined to ensure that they do a better job than fecal coliform in reflecting health risks (American Academy of Microbiology, 2001).

3.5 Full-Scale Facilities

According to an U.S. EPA in-house survey, there were 886 facilities treating CSOs in the United States in 1998 (Stinson, Wojtenko, and Field, 1999). Only about 20% of these facilities disinfected the CSO before discharging into the receiving waters (see Figure 3-1). Of the 20% of facilities disinfecting, the majority use gaseous chlorine as the disinfecting agent (108 out of 174 facilities). Over the last several years, there has been an increase in the number of CSO facilities changing to liquid instead of gaseous chlorine process. In addition, more facilities are also introducing the process of dechlorination in order to meet regulations and minimize harmful residuals in the receiving waters. According to the aforementioned in-house survey, 18 out of the 174 facilities were dechlorinating in addition to disinfecting (see Figure 3-1).

Since 1998, many new wet weather flow disinfection facilities have been constructed, most of them CSO chlorination facilities. However, disinfection of SW from separately sewered areas is gaining popularity, especially for beach communities in Southern California. UV disinfection for wet weather flows is also gaining popularity. For example, a UV disinfection facility for CSO treatment was constructed in the city of Bremerton, WA in 2001. A UV disinfection facility for SW treatment was also constructed at Moonlight State Beach in northern San Diego County in 2002.

Figure 3-1. Types of CSO Facilities in 1998 (Source: Stinson, Wojtenko and Field, 1999).

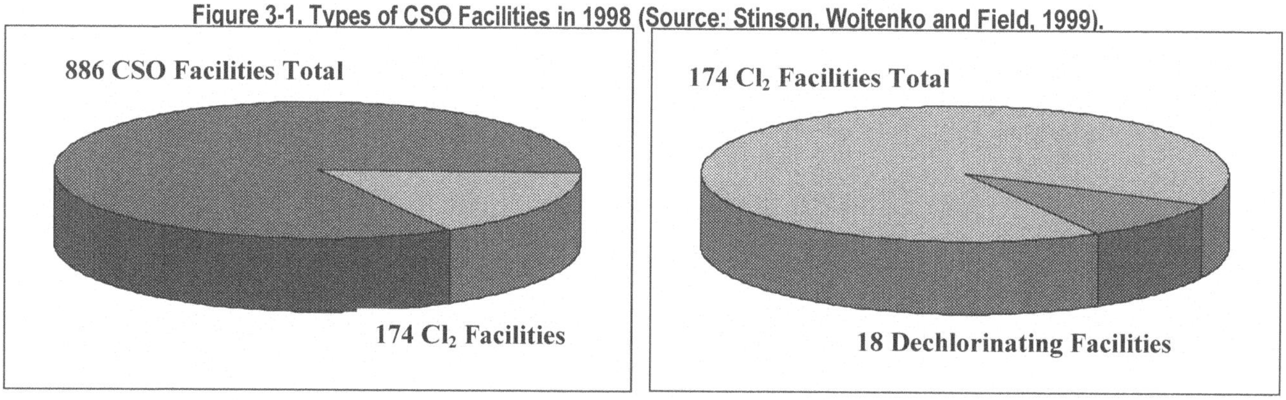

WERF

CHAPTER 4.0

DISINFECTION DEMONSTRATION

4.1 Disinfection Demonstration Scope of Work

The primary objective of the disinfection demonstration was to identify the benefits and potential risks associated with the disinfection of wet weather flows. This was accomplished by operating four available disinfection techniques (chlorination, chlorine dioxide, ozonation, and UV) and measuring their effectiveness in reducing indicator bacterial concentrations. During the operations, disinfection residuals and DBPs were measured to assess potential risks to aquatic life and human health.

The disinfection demonstration was a large bench-scale pilot test. Each pilot technology was sized to treat approximately 50–100 gallons of wet weather flow. Nine sampling events were conducted, each during a separate wet weather event. Each wet weather event exhibited different flow and water quality characteristics. During each of the nine test events, the dose of disinfectant was varied.

Wet weather flow was collected at the Metropolitan Sewage Treatment Plant (METRO), located in Syracuse, Onondaga County, New York, when influent flow was greater than 125 million gallons per day (MGD) or approximately two times the average dry weather flow (60 MGD). The METRO collection system is a combined sanitary and SW system. At METRO, wet weather flow was collected after the screening and degritting facilities but prior to primary clarification. This water quality was consistent with preliminarily-treated combined sewage such as flows that have been treated by high-rate physical treatment devices such as vortex technology.

Disinfectant dose versus bacterial reductions (dose-response) relationships were developed to gain an understanding of treatment effectiveness under different doses and water-quality characteristics. This effectiveness evaluation represented the benefits provided by disinfection. Measuring the concentrations of DBPs produced during the disinfection process portrayed the risks associated with disinfection. In addition, whole effluent toxicity (WET) tests and a DBP screening-level risk assessment were performed to portray the risks of disinfection.

For this disinfection demonstration, all three indicator bacterial organisms were tested (fecal coliform, *E. coli,* and *Enterococcus* spp.). The latter two indicators are considered to have a higher degree of association with gastrointestinal illnesses associated with swimming waters and were recommended by the U.S. EPA as the basis for bacterial water quality standards.

4.2 Disinfection Demonstration Methods and Materials

A detailed sampling plan, including equipment description, sampling protocol, analytical parameters, and quality assurance and quality control procedures, was developed before beginning the disinfection demonstration and was approved by the local stakeholders. The following is a summary description of the disinfection equipment, sampling protocol, and analytical parameters.

During each of nine storm events, 500 gallons of wet weather flow was collected and transported to a laboratory. At the laboratory, the wet weather flow was disinfected using the different disinfection technologies. Influent and effluent samples were collected to characterize the wet weather flow and the performance of the disinfection technologies.

One common influent water quality sample was collected at the beginning of each of the nine events, and the sample was analyzed for all parameters except bacteria. Bacterial samples were collected throughout the sampling because of the possibility of bacterial decay or growth during the sampling event. pH, specific conductivity (SpC), and temperature were remeasured when the bacterial samples were collected to determine if there was a change in water quality during the course of an event. Figure 4-1 depicts the sampling approach and the parameters that were measured.

4.2.1 Disinfection Equipment

The following is a list and description of the disinfection equipment and chemicals used.

Chlorination

Chlorine: 12% liquid sodium hypochlorite
Chlorine Dose: 8–28mg/L
Reducing Agent: 38% liquid sodium bisulfite
Reducing Agent Dose: assumed 55% chlorine dose remained as residual
Mixing/Contact Basin: 60-gal high-density polyethylene (HDPE) barrel
Mixing Power: ¼ hp per static 55-gallon volume
Contact Time: 5 min

Chlorine Dioxide

Chlorine Dioxide: acidified sodium chlorite
Chlorine Dioxide Dose: 5–15mg/L
Reducing Agent: solid ferrous sulfate (used to remove chlorite, not residual chlorine dioxide)
Reducing Agent Dose: 300–500 mg/L
Mixing/Contact Basin: 60-gal HDPE barrel
Mixing Power: ¼ hp per static 55-gal volume
Contact Time: 5 min

Ozonation

Ozonation: ozone generation by corona discharge process (Novaaqua Inc., Toronto, Canada)
Ozone Dose: 5–35mg/L
Flow Rate: 3–10 GPM
Mixing Power: 2-inch Mazi venturi eductor
Contact Basin: 500-ft of 1-in. diameter hose (rolled)
Contact Time: 3–5 min

UV

UV Lamp: medium pressure high intensity
Flow Rate: 3–10 GPM
UV Configuration: one lamp in stainless steal reactor chamber
UV Dose: 65–220 mW-sec/cm^2
Contact Time: less than 5 sec

4.3 Disinfection Demonstration Results

4.3.1 Event Summary

Nine individual storm events were sampled for the purpose of this disinfection demonstration. Each of the nine storm events met the flow and rainfall criteria for a wet weather event as was approved before starting the disinfection demonstration. Nine events were selected as a minimum number of events required to develop a disinfectant dose versus bacterial reduction (dose-response) relationship.

The proposed schedule for performing the disinfection demonstration started in June and ended in October of 2001. The demonstration began in July of 2001 but, due to lack of rain events, only four sampling events were completed by October 2001. The demonstration was shut down for the winter months and reconvened in April of 2002. The remaining five demonstration events were completed between April and June of 2002.

Table 4-1 presents the dates and times of each sampling event and the maximum METRO flow rate during sampling. The nine events varied in both rain intensities and flow rates. The events during the fall of 2001 tended to have greater rain intensities than the spring 2002 events; however, the flow rates tended to be greater during the spring 2002 events. This was likely the result of greater infiltration due to high groundwater during the spring months.

Table 4-1. Disinfection Demonstration Sampling Event Statistics

Event	Date	Sampling Start Time	Sampling End Time	Average Flow Rate During Sampling (MGD)	Rainfall Volume Preceeding Sampling Start Time (in)
1	July 17, 2001	9:30 AM	10:15AM	115	0.59
2	August 9, 2001	5:45 PM	6:30 PM	130	0.99
3	August 20, 2001	10:45 PM	11:45 PM	160	0.87
4	September 24, 2001	8:10 PM	9:05 PM	175	3.52
5	April 28, 2002	1:15 PM	2:15 PM	160	0.41
6	May 2, 2002	10:15 AM	11:15 AM	120	0.21
7	May 9, 2002	4:15 AM	5:35 AM	145	0.31
8	May 12, 2002	11:15 AM	12:30 PM	130	0.35
9	May 30, 2002	9:50 AM	10:50 AM	120	0.56

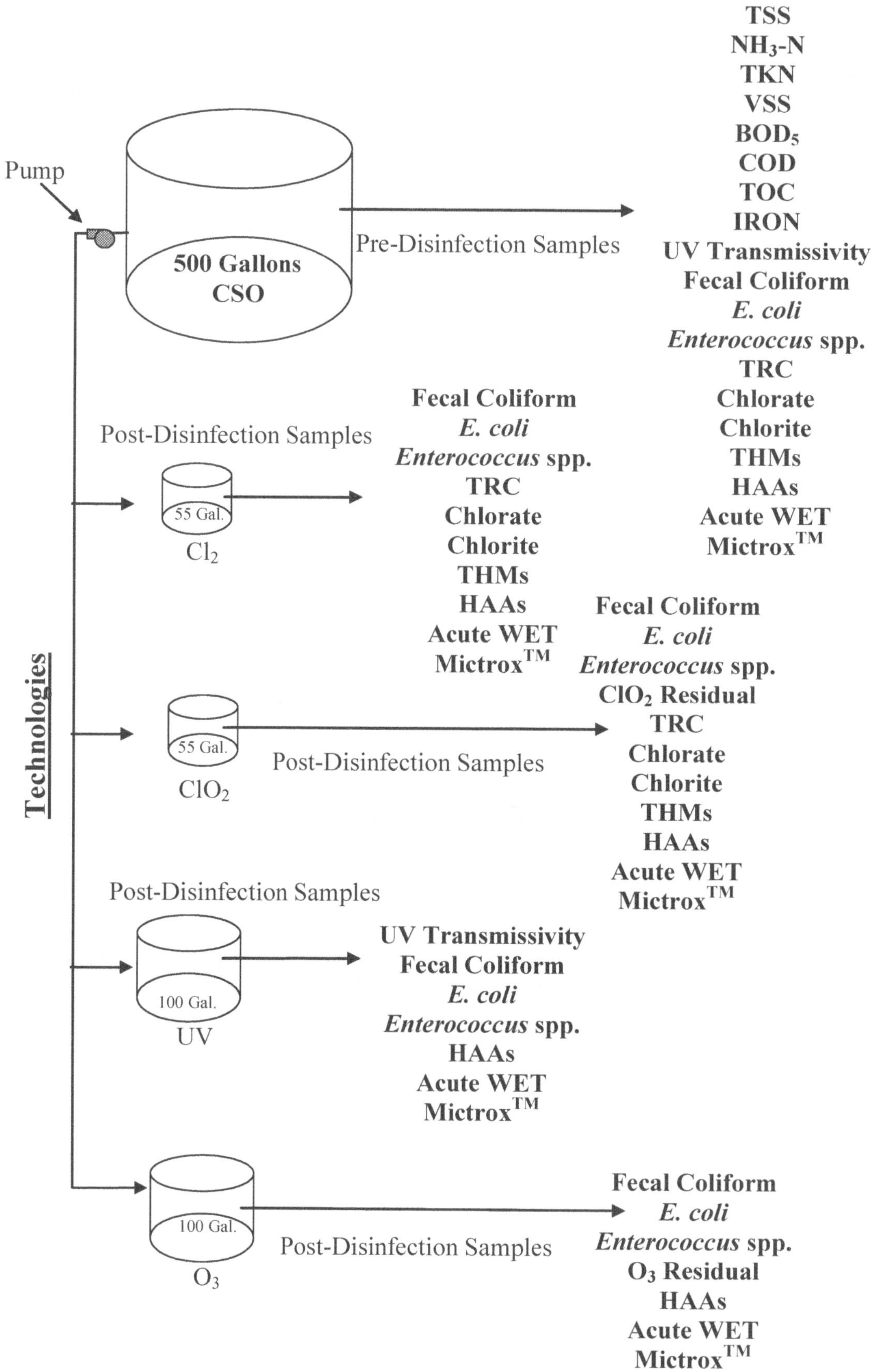

Figure 4-1. Disinfection Demonstration Sampling Approach.

4.3.2 Pilot Influent

The project team collected and tested wet weather flow from nine storm events. Common influent water quality samples were collected prior to operating the first disinfection pilot during each event. Because the bacteria parameters were the most important, and there was a possibility of bacterial decay or growth during a sampling event, l samples were collected from the influent just prior to operating each disinfection technology. pH, SpC, and temperature were re-measured when the bacterial samples were collected to determine if there was a change in water quality during the course of an event. The raw influent data and the calculated coefficients of determination (R^2) values among all influent parameters are provided in Tables 4-2 and 4-3, respectively. The R^2 values confirm relationships between influent parameters for the purpose of data analysis and quality assurance. Observations derived from the raw influent data are provided below.

The events may be segregated by water temperature into three categories: four events had water temperatures between 12°C and 13°C, another four events had temperatures between 18°C and 22°C, and one event had a water temperature of 26°C. Temperature is significant because it strongly correlates with fecal coliform and *E. coli* (R^2 values of 0.81 and 0.71, respectively); as the water temperatures increase, so do the bacterial concentrations. Temperature is also related to SpC (indirectly), transmissivity (indirectly), and iron.

The water quality of the wet weather flow samples may be characterized as a "weak" wastewater. The average biochemical oxygen demand (BOD) concentration was 76 mg/L; chemical oxygen demand (COD) averaged 182 mg/L, total organic carbon (TOC) averaged 64 mg/L, and total suspended solids (TSS) averaged 101 mg/L.

Events one and two have higher pollutant concentrations than the other seven events. These events have higher than average TSS, volatile suspended solids (VSS), BOD, COD, and TOC concentrations. The events also have some of the highest bacterial concentrations of the nine events.

TSS, COD, BOD, transmissivity, and *E. coli* show significant trends between each other indicating similar measurements of general water quality. TSS levels ranged between 29 and 200 mg/L.

Transmissivity also exhibited significant trends with temperature and fecal coliform.

Iron, a reducing agent, showed only general trends with bacterial concentrations and temperature.

No correlation was found between *Enterococcus* spp. and any of the other measured water quality parameters, including fecal coliform and *E. coli*. The strongest trends were with COD and iron.

Bacterial concentrations were typical of a CSO. Fecal coliform concentrations ranged from $1.0x10^5$ to $3.5x10^7$ CFU/100 mL, *E. coli* concentrations ranged from $3.0x10^4$ to $2.2x10^7$ CFU/100 mL, and *Enterococcus* spp. concentrations ranged from $2.9x10^3$ to $7.2x10^6$. The bacterial concentrations of the four cool-temperature events were generally one to two orders-of-magnitude less than the warmer temperature events.

Relationships between fecal coliform, *E. coli*, and *Enterococcus* spp. in the influent are shown in Figure 4-2. The correlation between fecal coliform and *E. coli* is strong, whereas the correlation with and between the other indicators are weak, showing only general trends. These data indicate that at this sampling location, historical fecal coliform data can be transformed into *E. coli* data with some certainty. Such transformation would not be appropriate from fecal coliform to *Enterococcus* spp. and between *E. coli* and *Enterococcus* spp.

Relationships between bacterial types are examined for two reasons. First, the U.S. EPA, in their new guidance on ambient water quality standards, recommends that states and tribes change from fecal coliform to *E. coli* and *Enterococcus* spp. because these indicators have better correlation with gastrointestinal illness rates for freshwater and marine water criteria, respectively. In accomplishing this, a better understanding of the relationship between the different organisms in untreated and treated water is important. The second reason is to show the differences in performance on reducing specific organisms for the different disinfection processes. These differences in bacterial relationships will be pointed out in subsequent discussions under each disinfection process.

Influent water was not toxic to fathead minnows for each of the first four sampling events when this assay was performed. In each instance there was no mortality observed among minnows exposed to the whole influent sample (i.e., no dilution) for a period of 48 hours in preliminary screening tests. Influent water was also not toxic to the invertebrate *Ceriodaphnia dubia* for each of the nine sampling events in this study. There was an insignificant level of mortality (<15%) among *C. dubia* exposed to whole influent after an exposure period of 48 hours in preliminary screening tests. Consequently, the reported median lethal concentration (LC_{50}) for each species for each collection event was >100%.

Microtox bacterial cultures exhibited minor reductions in luminescence on exposure to whole influent sample in each of the nine sampling events. Changes in luminescence in multi-replicate inhibition tests ranged from slight increases relative to control levels to no more than a 50% reduction in light output. Consequently, the reported 15-minute median effective concentrations (EC_{50}) for influent from each collection event was >100%.

Table 4-2. Disinfection Demonstration Influent Water Quality Data.

Event No.	1	2	3	4	5	6	7	8	9	AVG	STD
Environmental Data											
TSS (mg/l)	160	200	29	74	89	77	68	91	120	101	51.8
VSS (mg/l)	120	130	20	50	63	50	43	62	91	70	36.5
BOD (mg/l)	88	90	49	61	85	61	75	86	91	76	15.5
Transmissivity (%)	54	39	62	64	63	70	65	66	66	61	9.3
Ammonia (mg/l)	8	5.7	4.3	4.9	8.6	7.7	4.7	9.5	10	7	2.2
TKN (mg/l)	16	11	8.2	9.6	14	14	9.7	17	19	13	3.7
COD (mg/l)	280	300	140	130	180	110	140	160	200	182	66.9
TOC (mg/l)	74	76	50	45	65	62	48	55	100	64	17.5
Iron (mg/l)	8	6	5.7	5.8	4.3	2.5	4	3.5	6.3	5.1	1.7
Dissolved Oxygen (mg/l)	1.8	1.3	2.7	1.2	1.1	0	1.3	1.6	2.5	1.5	0.8
pH	6.98	6.75	6.95	7.35	6.94	7.4	7.1	8.4	7.17	7	0.5
Spec. Conductance (μmhos/cm)	850	590	580	600	1360	2000	980	1450	1280	1,077	485.4
Temperature (°C)	20	26	21.5	20	12	12	13	12	18	17	5.1
Bacteria											
Fecal Coliform (cfu/100ml)	1.3E+07	3.5E+07	9.0E+06	1.5E+07	4.2E+05	2.8E+05	4.0E+05	4.8E+05	4.1E+06	8.7E+06	1.1E+07
E. coli (cfu/100ml)	1.2E+07	1.5E+07	4.0E+06	5.0E+06	6.0E+04	7.5E+04	6.2E+04	1.1E+05	8.7E+05	4.1E+06	5.7E+06
Enterococcus spp. (cfu/100ml)	4.6E+06	3.0E+05	1.3E+05	2.1E+05	7.7E+03	3.8E+03	2.8E+04	2.8E+05	3.4E+05	6.5E+05	1.5E+06
Disinfection By-Products											
Chlorate (mg/l)	0.051	<0.02	0.09	0.09	<0.05	<0.05	<0.025	<0.025	<0.025	na	na
Chlorite (mg/l)	0.043	<0.02	<0.02	<0.02	<0.05	<0.05	<0.025	<0.025	<0.025	na	na
Total Residual Chlorine (mg/l)	<0.03	<0.03	<0.03	<0.03	<0.03	<0.03	<0.03	<0.03	<0.03	na	na
Residual Chlorine Dioxide (mg/l)	<0.03	<0.03	<0.03	<0.03	<0.03	<0.03	<0.03	<0.03	<0.03	na	na
Ozone Residual (mg/l)	<0.03	<0.03	<0.03	<0.03	<0.03	<0.03	<0.03	<0.03	<0.03	na	na
Formaldehyde (mg/l)	nm	<1	1.4	nm	nm	nm	nm	nm	nm		
THMs											
Chloroform (μg/l)	2	2	1.6	2	2	1	<1	3	3	2	0.7
Bromodichloromethane (μg/l)	<1	<1	<0.5	<1	<1	<1	<1	<1	<1	na	na
Dibromochloromethane (μg/l)	<1	<1	<0.5	<1	<1	<1	<1	<1	<1	na	na
Bromoform (μg/l)	<1	<1	<0.5	<1	<1	<1	<1	<1	<1	na	na
HAAs											
Dibromoacetic acid (μg/l)	<1	<1	<1	<1	<1	<1	<1	<1	<1	na	na
Dichloroacetic acid (μg/l)	1.4	6.1	4.7	3.8	5.7	5	3.1	7.4	5.9	5	1.8
Monobromoacetic acid (μg/l)	<1	<1	<1	<1	<1	<1	<1	<1	<1	na	na
Monochloroacetic acid (μg/l)	<2	3.9	<2	<2	<2	<2	<2	<2	<2	na	na
Trichloroacetic acid (μg/l)	3.1	6.4	29	3.6	8.1	8.3	2.7	6.3	6	8	8.1
Total Haloacetic Acid (μg/l)	8.5	18	38	11	14	13	5.8	14	12	15	9.3
Toxicity											
Acute WET Minnow (LC50-%)	>100	>100	>100	>100	nm	nm	nm	nm	nm	>100	na
Acute WET Water Flea (LC50-%)	>100	>100	>100	>100	>100	>100	>100	>100	>100	>100	na
Microtox™ (EC50-%)	>100	>100	>100	>100	>100	>100	>100	>100	>100	>100	na

Table 4-3. Disinfection Demonstration Influent Water Quality R^2 Comparison.

	TSS	VSS	BOD	Trans	NH3	TKN	COD	TOC	Iron	DO	pH	SpC	Temp	FC	*E. coli*	*Entero*
TSS	**1.00**															
VSS	**0.98**	**1.00**														
BOD	**0.58**	**0.63**	**1.00**													
Trans	**0.63**	**0.55**	0.16	**1.00**												
NH3	0.08	0.13	0.42	0.06	**1.00**											
TKN	0.13	0.20	0.46	0.04	**0.95**	**1.00**										
COD	**0.85**	**0.87**	**0.50**	**0.75**	0.03	0.01	**1.00**									
TOC	0.41	**0.50**	0.45	0.07	0.41	**0.50**	0.36	**1.00**								
Iron	0.24	0.32	0.07	0.30	0.01	0.00	0.47	0.17	**1.00**							
DO	0.00	0.00	0.01	0.01	0.00	0.01	0.06	0.07	0.38	**1.00**						
pH	0.07	0.07	0.26	0.26	0.18	0.16	0.17	0.07	0.25	0.02	**1.00**					
SpC	0.03	0.03	0.36	0.36	0.44	0.34	0.15	0.03	**0.50**	0.29	0.22	**1.00**				
Temp	0.24	0.22	**0.60**	**0.60**	0.19	0.11	0.37	0.05	**0.57**	0.19	0.29	**0.65**	**1.00**			
FC	0.47	0.39	**0.84**	**0.84**	0.15	0.11	0.49	0.03	0.32	0.01	0.21	0.46	**0.81**	**1.00**		
E. coli	**0.58**	**0.54**	**0.85**	**0.85**	0.08	0.03	**0.69**	0.05	**0.50**	0.02	0.23	0.41	**0.71**	**0.88**	**1.00**	
Entero	0.22	0.31	0.10	0.10	0.03	0.10	0.35	0.07	0.46	0.03	0.03	0.04	0.06	0.08	0.28	**1.00**

Legend: TSS=total suspended solids; VSS=volatile suspended solids, BOD=biochemical oxygen demand; Trans=transmissivity; NH3=ammonia; TKN=total Kjeldahl nitrogen; COD=chemical oxygen demand; TOC=total organic carbon, DO=dissolved oxygen; SpC=specific conductance; Temp=temperature; FC=fecal coliform; Entero=*Enterococcus* spp.

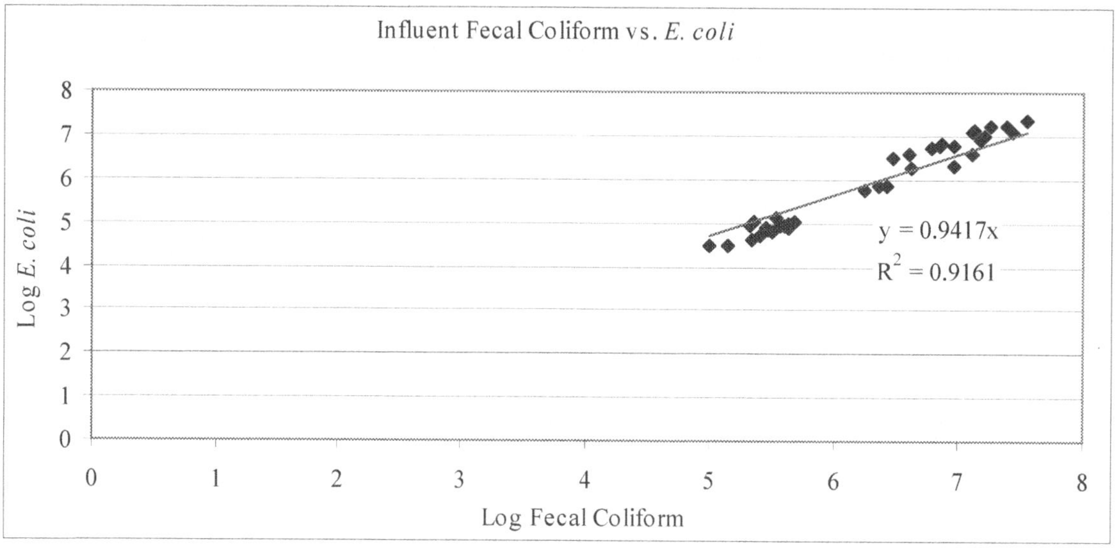

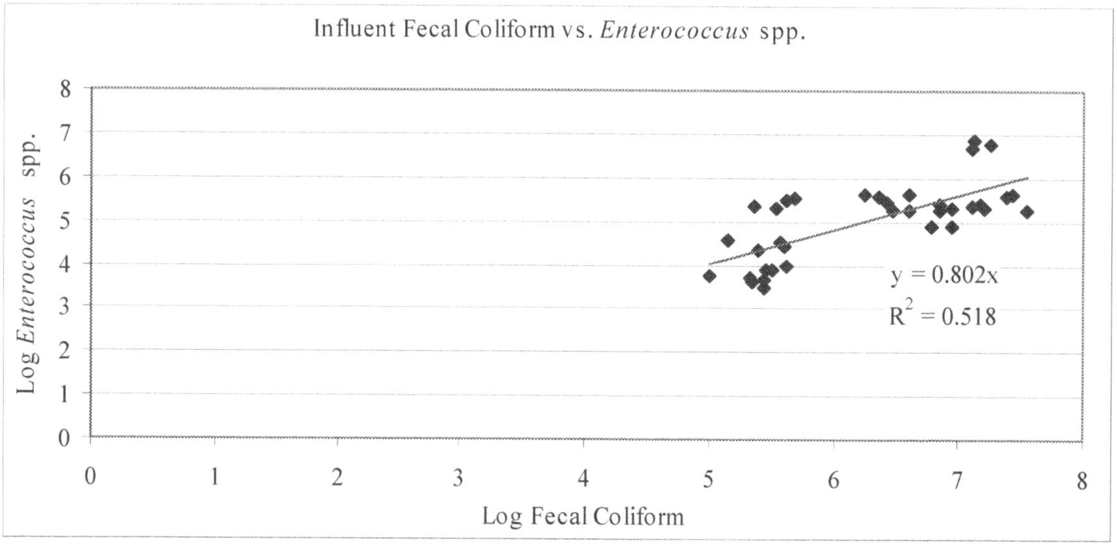

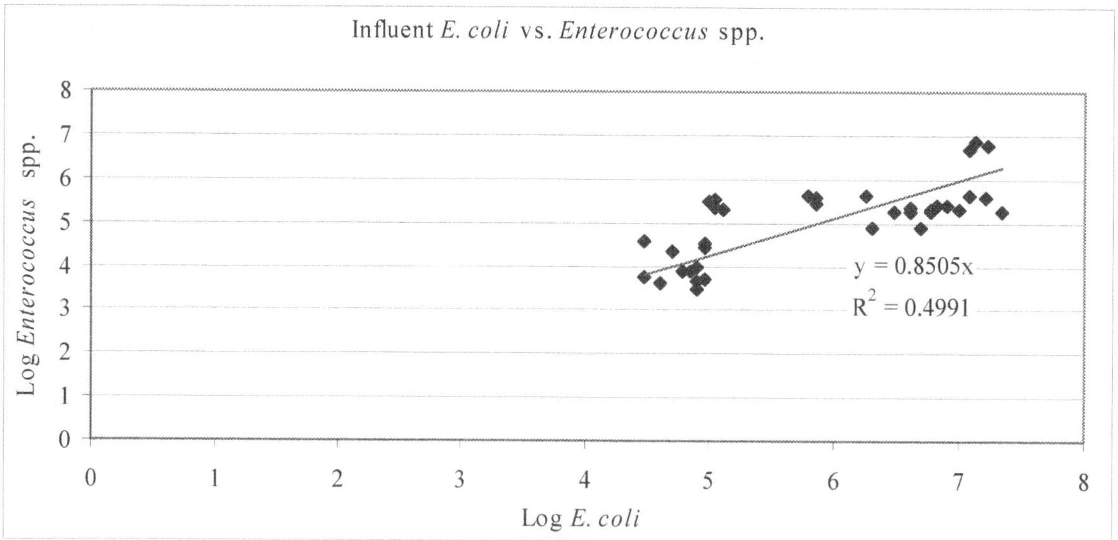

Figure 4-2. Comparison of the Indicator Bacterial Organisms in the Influent.

4.3.3 Chlorination

Chlorine disinfection was tested on nine wet weather flow samples with varying water quality. The dose concentration ranged from 8.5 to 28 mg/L. During each event, the chlorine was applied with a high rpm mixer, and five minutes of contact time was provided. The ranges of percent reduction for the three types of bacteria are summarized in Table 4-4. In terms of percent reduction, chlorine was most effective in inactivating fecal coliform and least effective in inactivating *E. coli*.

Table 4-4. Bacterial Data Summary for Chlorination Disinfection.

Bacterial Type	Initial Count Range (CFU/100 mL)	% Reduction Range	Avg. Log Reduction
Fecal coliform	2.2E5 to 3.5E7	99.960 to 99.998	3.3
E. coli	4.0E4 to 2.2E7	98.722 to 99.995	2.3
Enterococcus spp.	4.1E3 to 4.1E5	72.340 to 99.947	3.1

4.3.3.1 Dose-Response Relationships

Scatter plots were developed for chlorine dose versus log of effluent bacterial concentrations and log bacterial reductions for each of the three bacterial groups. Linear, logarithmic, and power regressions were tested against each data set, and the strongest relationship was used for calculating the coefficient of determination shown on each plot. The strongest relationship was usually observed using a power regression equation.

Previous disinfection studies by WWETCO (Boner and Turner, 2002) and others have shown that dose-response relationships may be improved by normalizing the applied dose by influent water quality. To determine if the dose-response curves could be improved, R^2 values were developed to test all possible combinations of dose normalized by one and two parameters. Several water quality constituents, including COD, TOC, TSS, BOD, and iron, were used to normalize the applied dose, but COD provided the best correlation factor. All of these dose-response relationships required temperature to be the second normalization parameter to ensure a significant correlation factor. Adding other normalization parameters such TSS and TKN to COD and temperature did not measurably improve the correlation factor. An example of the normalization process is shown in Figures 4-3 for fecal coliform. Figure 4-3 shows a series of charts with different X-axes, including applied dose, applied dose divided by COD, and applied dose divided by both COD and temperature.

Because COD and temperate provided the best correlation, only the dose-response curves normalized by COD and temperature were used for data analysis in Chapter 4.0. However, in Chapter 5.0, *Applications for Full-Scale Facilities*, dose-response curves were normalized by TSS and temperature for the purpose of developing an algorithm, which in turn can be used for process control with instrumentation such as TSS or turbidity meters.

Figure 4-3 is presented as an example of the normalization process for fecal coliform. The final result of the normalization process is the chart with the x-axis labeled "Dose Normalized by COD and Temperature." These final normalized chlorine dose-response curves were developed for each of the bacterial groups including fecal coliform, *E. coli*, and *Enterococcus* spp. and are presented in Figure 4-4.

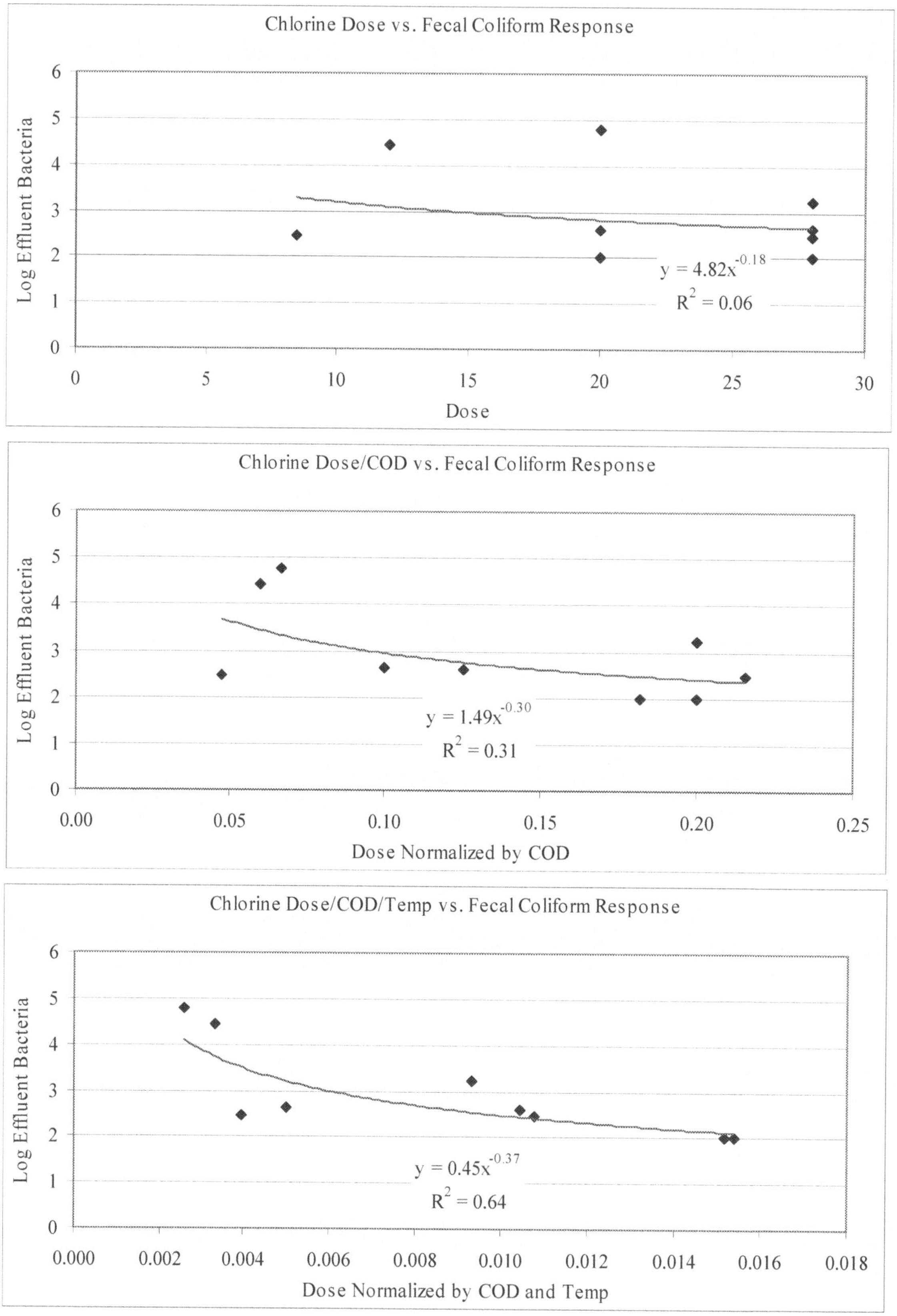

Figure 4-3. Example of Chlorine Dose Normalization Process for Fecal Coliform.

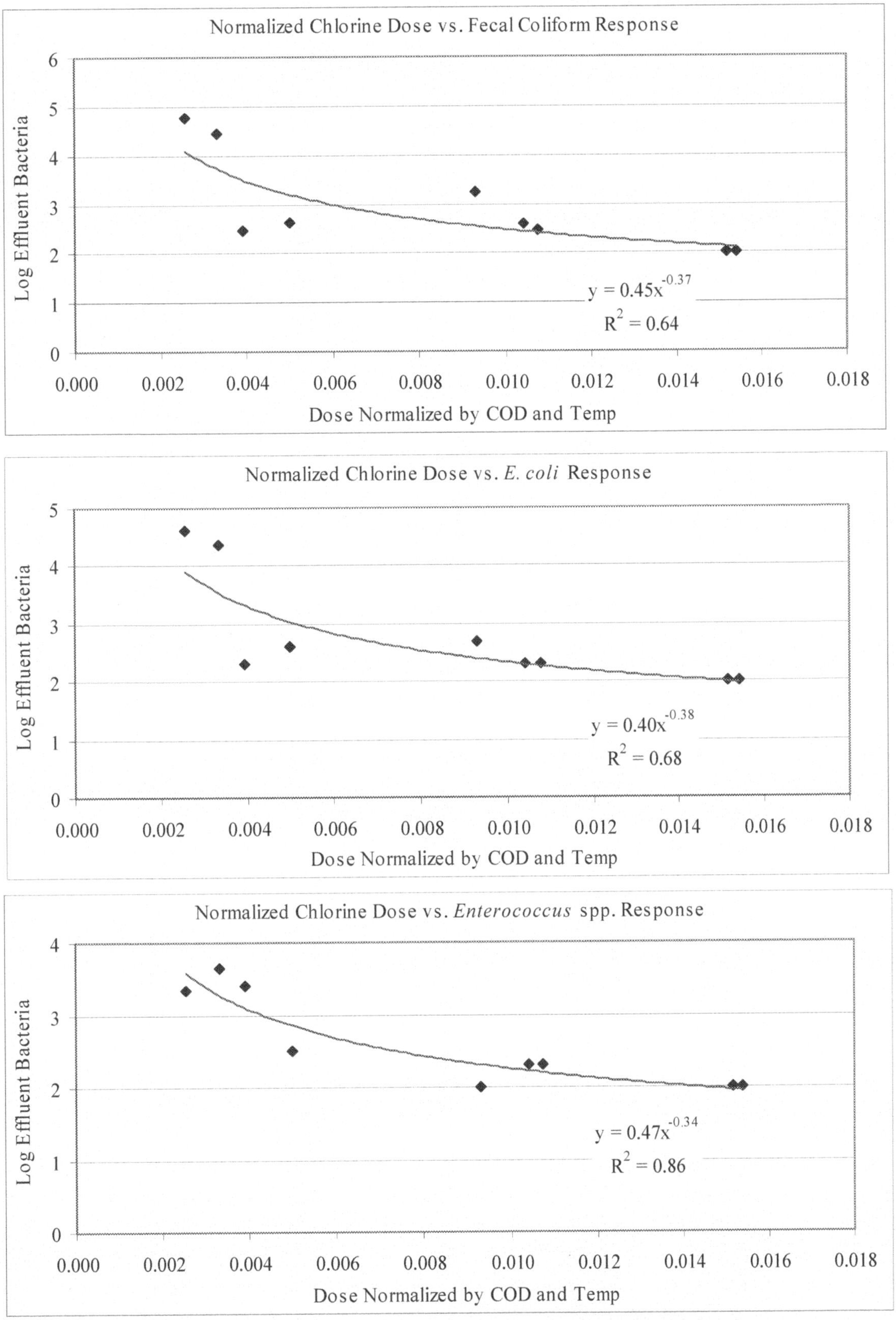

Figure 4-4. Chlorine Dose-Response Curves Normalized by COD/Temperature.

Table 4-5 summarizes the chlorine dose necessary to achieve the maximum value criteria and bacterial concentrations of 1,000 CFU/100 mL. A bacteria density of 1,000 CFU/100 mL was chosen as a benchmark to compare the technologies, and is a common effeluent limit for fecal coliform. The dose values are derived from the normalized dose-response curves, Figure 4-4. For this demonstration, typical COD and temperature values for wet weather flow at the METRO plant were assumed to be 200 mg/L and 20°C, respectively. Therefore, the normalized dose in Figure 4-4 was multiplied by 200 mg/L COD and 20°C to derive the dose needed to achieve the desired effluent bacterial concentration.

Table 4-5. Chlorine Dose Required to Meet Bacterial Criteria.

Parameter	Cl_2 Dose to Achieve Maximum Value Criteria	Maximum Value Criteria	Cl_2 Dose to Achieve 1,000 CFU/100 mL
Fecal coliform	18 mg/L	4,274 CFU/100 mL	25 mg/L
E. coli	18 mg/L	2,507 CFU/100 mL	22 mg/L
Enterococcus spp.	21 mg/L	656 CFU/100 mL	18 mg/L

Note: The above table uses the correlations based upon dose normalized by COD and temperature.

When considering the same criteria for each bacterial group (1,000 CFU/100 mL), fecal coliform required the highest dose. This does not indicate that fecal coliform was least susceptible to chlorine, but rather indicates that fecal coliform had the highest influent concentrations (i.e., highest challenge concentration). Conversely, the maximum bacterial criteria become more stringent in the order of fecal coliform, *E. coli*, and *Enterococcus* spp. and hence required greater disinfectant doses to achieve the stated maximum value criteria.

In addition to COD and temperature, disinfection efficiency of chlorine is significantly reduced by the production of chloramines, which can be produced by chemical reaction between chlorine and ammonia. For example, for fecal coliform, the biocide potency of $HOCl:OCl^-$ to $NH_2Cl:NHCl_2$ is approximately 1:0.0125 to 0.005:0.0166 (Chang, 1971). Hence, ammonia concentration is an important factor to consider when determining effectiveness of chlorine.

Chemical reaction rates for the formation of chloramines are a function of chlorine and ammonia concentrations and temperature. Chlorine will convert to mono- and dichloramines when the ratio of chlorine to ammonia is less than eight. When this ratio is greater than eight, most of the chlorine remains as free chlorine; this is breakpoint chlorination. Based on the ratio of applied chlorine dose to measured ammonia in this demonstration, it is certain that breakpoint chlorination did not occur, resulting in the formation of mono- and dichloramines. Nevertheless, chlorination, regardless of form, was effective at inactivating the indicator organisms.

Relationships between fecal coliform, *E. coli*, and *Enterococcus* spp. in the chlorine-disinfected effluent are shown in Figure 4-5. Similar to the influent data (Section 4.3.2), these relationships indicate that fecal coliform data from a chlorinated effluent can be transformed into *E. coli* with some level of certainty, but transformation would not be appropriate from fecal coliform to *Enterococcus* spp., and between *E. coli* and *Enterococcus* spp. Additionally, comparing the slopes of the lines from the influent charts (Figure 4-2) to the effluent charts (Figure 4-5) provides an indication of the effectiveness of chlorine. Based on changes to the slope of the lines, chlorine was most effective at inactivating *Enterococcus* spp., which was likely because fecal coliform and *E. coli* had the highest challenge concentrations.

4.3.3.2 Disinfection Residuals and DBPs

The concentration of disinfection residuals and DBPs, including total residual chlorine (TRC), THMs, and haloacetic acids (HAAs), were measured in the influent and effluent of the chlorine pilot for each of the sampling events. Effluent results are presented in Table 4-6.

Table 4-6. Chlorination/Dechlorination Effluent Disinfection Residual and DBP Data.

Event	1	2	3	4	5	6	7	8	9
Dose (mg/L)	28	20	28	28	8.5	20	12	20	12
Chlorate (mg/l)	6.9	5.3	8.6	6.1	7.7	1.7	<0.25	2.4	1.21
Chlorite (mg/l)	0.042	<0.02	<0.02	<0.2	<0.5	<0.5	<0.25	<0.25	<0.25
TRC - pre-dechloriantion (mg/l)	6.1	2.2	2.6	4.95	0.46	13.9	6.76	15.9	6.46
TRC - post-dechloriantion (mg/l)	<0.03	<0.03	<0.03	<0.03	<0.03	5.2	2.46	5	2.62[(1)]
Chloroform (ug/l)	6	8	6.9	6	4	6	7	6	5
Bromodichloromethane (ug/l)	<1	<1	<0.5	<1	<1	<1	<1	<1	<1
Dibromochloromethane (ug/l)	<1	<1	<0.5	<1	<1	<1	<1	<1	<1
Bromoform (ug/l)	<1	<1	<0.5	<1	<1	<1	<1	<1	<1
Dibromoacetic acid (ug/l)	<1	<1	<1	<1	<1	<1	<1	1.5	<1
Dichloroacetic acid (ug/l)	14	19	17	8.1	12	22	23	24	17
Monobromoacetic acid (ug/l)	<1	<1	<1	<1	<1	<1	<1	<1	<1
Monochloroacetic acid (ug/l)	<2	3.6	<2	<2	2.4	2.5	2.4	3.5	<2
Trichloroacetic acid (ug/l)	3.2	6.7	36	6.1	8.2	15	8.3	11	8.3
Acute Toxicity - Fathead Minnow	>100	>100	>100	>100	ND	<10	ND	ND	ND
Acute Toxicity - *C. dubia*	>100	>100	>100	>100	>100	<7	<14	<14	>14<71
Microtox™	>100	>100	>100	>100	>100	7.7	11.8	4.9	29.9

All results are from post-dechlorination samples except for "TRC – pre-dechlorination".(1) TRC after "low dechlorination dose".

Similar to the dose-response curves, disinfection residuals and DBPs were plotted against dose and dose normalized by COD and temperature. The curve that provided the best correlation was used to quantify the level of residuals and DBPs found in the effluent.

Total residual chlorine is often a measure of the residual from the chlorine disinfection process. It was found that TRC (pre-dechlorination) was a function of the applied dose as shown in Figure 4-6. In order to use Figure 4-6 to predict a TRC, an applied dose must be selected and values for COD and temperature assumed. Based on the finding presented in Table 4-5, an applied dose of 18 mg/L was selected as the dose to achieve the maximum value criterion for *E. coli*. Additionally, for this demonstration, typical COD and temperature values for wet weather flow at the METRO plant were 200 mg/L and 20°C, respectively. Therefore, based on an applied dose of 18 mg/L, a COD of 200 mg/L and a temperature of 20°C, TRC before dechlorination can be estimated from Figure 4-6 to be approximately 4 mg/L (dose of 18 divided by 200 COD and 20°C temperature equals a normalized dose of 0.0045).

TRC concentrations (post-dechlorination) were below the detection limit of 0.03 mg/L during events one through five. However, during events six through nine, all in the spring of 2002, TRC values between 2.5 mg/L and 5.2 mg/L were measured in the effluent after dechlorination because insufficient quantities of sodium bisulfite were added. For each event, it was assumed that TRC in the effluent would be equivalent to 55% of the applied dose for the purpose of adding sodium bisulfite. During events six through nine, the 55% residual assumption did not hold true due to "weaker" wastewater quality during the spring runoff season. The

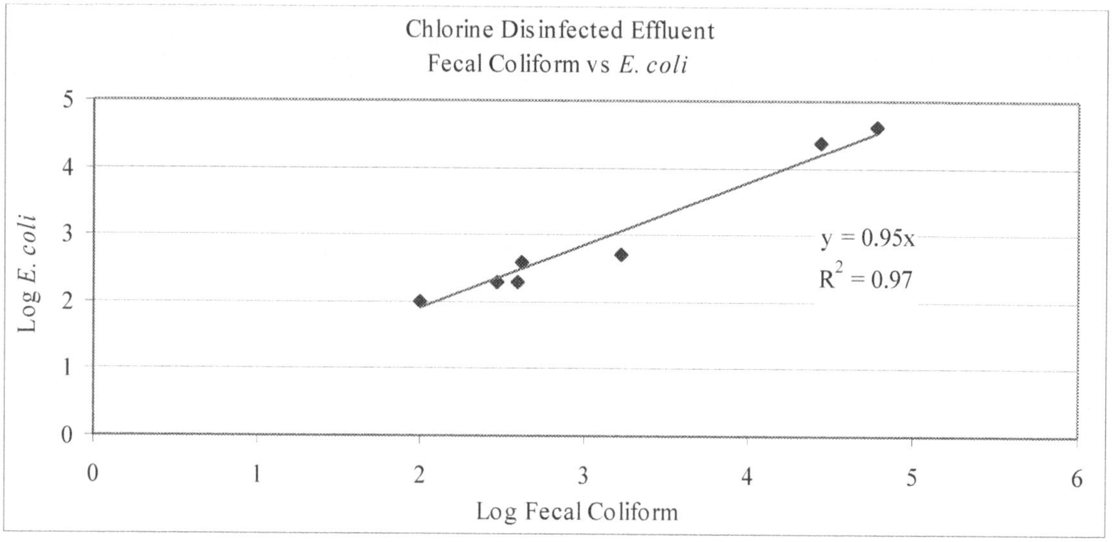

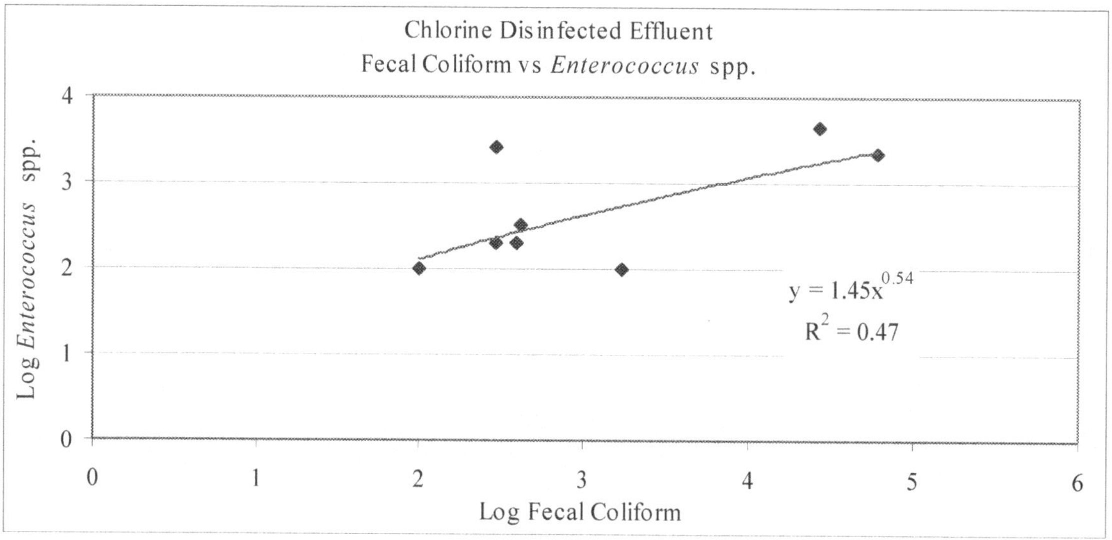

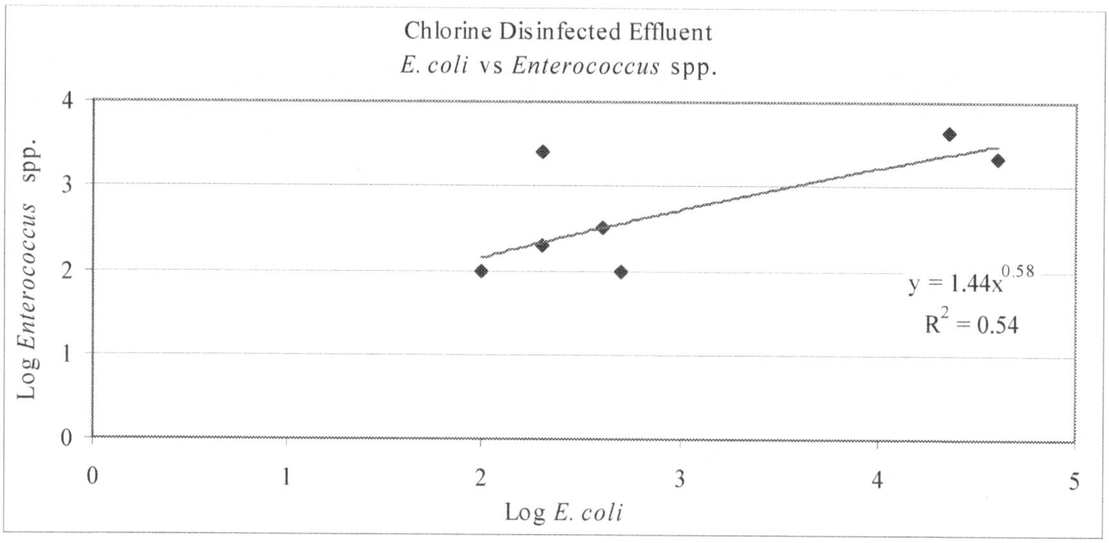

Figure 4-5. Comparison of the Indicator Bacterial Organisms in the Chlorine Effluent.

chlorine demand was low and therefore chlorine residual was greater than 55%, resulting in TRC being present in the effluent after dechlorination.

Microtox assays were performed only on post-dechlorination samples. No significant toxicity was observed in the samples collected during events one through five when no TRC was present. Conversely, significant toxicity was observed during events six through nine when TRC was present. Post-dechlorination TRC was compared to the Microtox EC_{50} as shown in Figure 4-7. This plot indicates that any measurable TRC will cause increased aquatic toxicity.

As presented in Figure 4-6, an applied dose of 18 mg/L is expected to result in an effluent TRC of 4 mg/L before dechlorination. Based on the TRC to aquatic toxicity relationship presented in Figure 4-7, it is expected that a TRC of 4 mg/L would cause significant toxicity if aquatic organisms were exposed to the undiluted effluent. Therefore wet weather flows discharging into sensitive receiving waters will likely require dechlorination to minimize aquatic toxicity. However, this determination should be made by examining the dynamics of the discharge and receiving waterbody and should consider allowable mixing zones.

For event nine, sodium bisulfite was added per the 55% TRC assumption, which resulted in a post-dechlorination TRC of 2.6 mg/L in the effluent. Toxicity analyses were run on this effluent sample, which resulted in significant toxicity to the aquatic test organisms. An additional dose of sodium bisulfite was added to the effluent to reduce the TRC to less than 0.03 mg/L. Toxicity analyses were also run on this effluent sample resulting in no toxicity to the aquatic test organisms. The toxicity results are discussed in more detail in the Section 4.3.3.3. This exercise illustrates the importance of proper residual chlorine control through dechlorination.

Chlorates were consistently formed in the chlorination disinfection process. Effluent chlorate concentrations compared to doses are shown in Figure 4-8. This relationship indicates that chlorates are produced in linear proportion to applied doses greater than approximately 10 mg/L chlorine. Toxicology information on chlorate is limited, with only acute chlorate toxicity having been addressed (White, 1999). For drinking water, the U.S. EPA's Safe Drinking Water Act (SDWA) does not set a limit on chlorate but suggests that its production be limited. Toxicity issues related to chlorates and other DBPs are discussed in more detail in Section 4.3.3.3.

Of the THMs measured in the disinfection demonstration, chloroform is the only DBP that appeared to be produced in the chlorination disinfection process. Dose normalized by COD and temperature is compared to effluent chloroform concentration as shown in Figure 4-9. The correlation shows a limited power trend; chloroform is produced as dose is increased but at a slightly reduced rate. It is generally understood that chlorination produces THMs if the precursors are available. The average chloroform concentration was 6.1 μg/L; this is an approximate three-fold increase over influent concentrations. Chloroform is regulated by the U.S. EPA under the SDWA but has no maximum contamination level goal (MCLG). A proposed MCLG for chloroform is 70 μg/L.

Of the HAAs, dichloro- and trichloroacetic acids were the only forms produced in the chlorination disinfection process. Dichloro- and trichloroacetic acids were plotted against the dose normalized by COD and temperature as shown in Figures 4-10 and 4-11, respectively. The plots show a general trend that indicates these substances are produced in linear proportion to the normalized dose. On average, the total HAA effluent concentration was approximately two times the influent concentrations. The total HAA effluent concentrations ranged between 18 and 57 μg/L. The SDWA maximum contamination level (MCL) for total HAA is 60 μg/L.

A summary of TRC and DBPs is presented in Table 4-7. The values in this table were derived for the typical wet weather flow quality at METRO, 200 mg/L COD and 20°C, and an applied dose of 18 mg/L as required to achieve the maximum value *E. coli* criterion. The TRC concentration and resulting aquatic toxicity reflect the chlorinated effluent before dechlorination.

Table 4-7. Chlorination Residuals and DBPs for Typical Wet Weather Flow at METRO.

Parameter	Value
Dose	18 mg/L
TRC	4.3 mg/L
Microtox (EC_{50}%)	56%
Chlorate	3.0 mg/L
THM-Chloroform	4.0 μg/L
HAA-Dichloroacetic acid	19 μg/L
HAA-Trichloroacetic acid	10 μg/L

Note: The dose in the above table is the required dose to obtain an *E. coli* concentration of 2,507 CFU/100 mL (Table 4-5). TRC is before dechlorination, and the resulting Microtox is based on 4.3 mg/L TRC. It is important to note that if sufficient dechlorination agent is added, then TRC would be less than 0.03 mg/L and no aquatic toxicity would be observed.

4.3.3.3 Toxicity

Measured Aquatic Toxicity

Chlorinated/dechlorinated effluent was not toxic to fathead minnows during events when this assay was performed (events one through four). In each instance there was no mortality observed among minnows exposed to whole treated sample for a period of 48 hours in preliminary screening tests. Chlorinated/dechlorinated effluent was also not toxic to the invertebrate, *C. dubia*, for events one through five (LC_{50}>100%), which was in contrast with significant levels of mortalities for the remaining events. All *C. dubia* exposed to 10% and 20% dilutions of whole sample in events six and seven, respectively, died within 24 hours resulting in estimated an LC_{50} of <7 and <14%, respectively. Results from event eight indicated that the LC_{50} would fall in the range from 2.8% to 14% of the whole sample. The toxicity of two separate samples was assessed for the last event (nine) in which differing dechlorination levels were applied to chlorinated sample. The low dechlorinated sample, which resulted in a TRC of 2.6 mg/L, exhibited a significant level of toxicity with an LC_{50} ranging from 14% to 71%. The high dechlorinated sample, which resulted in a TRC of <0.03 mg/L, exhibited no toxicity (LC_{50}>100%).

Microtox bacterial cultures exhibited minor reductions in luminescence on exposure to chlorinated/dechlorinated sample in each of the first five demonstration events. Samples from events six through eight were all sufficiently toxic to generate EC_{50} values ranging from 4.9% to 12%. Results from event nine were similar to those obtained with *C. dubia*. The low dechlorinated sample (TRC = 2.6 mg/L) was toxic generating an EC_{50} of 30%. The high dechlorinated sample (TRC < 0.03 mg/L) resulted in an EC_{50} > 100%.

Literature Review Toxicity Findings

A summary of literature review toxicity data for chlorine residuals, THMs, and HAAs is presented in Table 4-8. Table 4-8 only shows residuals and DBPs that increased in concentration as a result of the chlorination process. Additional information on these chemicals and other chemicals can be found in Chapter 8.0. Table 4-8 presents acute and chronic toxicity values for fresh (FW) and marine (SW) water animals; the circled values are the lowest or most conservative values found. Mammalian toxicity reference doses (RFD) for chlorine, chloroform, monochloroacetic acid, and dichloroacetic acid were found, as well as cancer slope factors (CSF) for chloroform and dichloroacetic acid, but reference values for the other chemicals were not found to be published. Chlorine residuals, THMs, and HAAs may cause cancer in mammalian species and affect reproductive organisms, but the current data base is inconclusive. Some organs affected by these chemicals are the bladder, liver, and kidney.

Table 4-8. Toxicity Literature Review Findings for Chlorination.

		Acute Toxicity Value			Chronic Toxicity Value			Mammalian Toxicity			
Chemical	**CAS NO**	**Species**	**Value (µg/L)**	**Endpoint**	**Species**	**Value (µg/L)**	**Endpoint**	**Carcinogen**	**Possible Organs Affected**	**RfD (mg/Kg-day)**	**CSF $(mg/Kg\text{-}day)^{-1}$**
Total Chlorine Residual (TCR)	7790-92-3	FW Fish: FW Invert: SW Fish: SW Invert:	(100) ND ND ND	2-hr LC50 ND ND ND	ND			Possibly	bladder	$1x10^{-1}$	ND
Chlorate	14866-68-3	ND			ND			ND	ND	ND	ND
Chloroform	67-66-3	FW Fish: FW Invert: SW Fish: SW Invert:	(13300) 2000 ND 2400	96-hr LC50 1-hr LC50 ND 1-hr LC50	FW Fish: FW Fish: FW Invert: FW Invert: SW Fish: SW Invert:	1240 122000 235000 6300 ND ND	672-hr LC50 240-hr NOEC 216-hr LC50 504-hr NOEC ND ND	Probable	bladder, colon, liver, kidney	$1x10^{-2}$	RfD is health protective for cancer
Monochloroacetic acid (metabolite of DCA)	79-11-8	FW Fish: FW Invert: SW Fish: SW Invert:	ND (77000) ND ND	ND 48-hr EC50 ND ND	FW Fish: FW Invert: SW Fish: SW Invert:	ND 32000 ND ND	ND 504-hr NOEC ND ND	ND	liver, testicle, heart, vessels, kidney	0.002 (1)	ND
Dichloroacetic acid (DCA)	79-43-6	FW Fish: FW Invert: SW Fish: SW Invert:	ND ND ND (23000)	ND ND ND 96-hr LC50	ND			Probable	liver, testicle, heart, vessels, kidney	$4x10^{-3}$	$5x10^{-2}$
Trichloroacetic acid	76-03-9	FW Fish: FW Invert: SW Fish: SW Invert:	1050000 (1200) 9300000 4800000	96-hr LC50 24-hr LC50 96-hr LC50 96-hr LC50	ND			Possible	liver, testicle, heart, vessels, kidney	ND	ND

Circled value is the lowest available ecotoxicity value and is therefore used as the screening toxicity value for the risk assessment.
ND = No Data or Insufficient Data
RfD = Reference Dose
CSF = Cancer Slope Factor
(1) This reference dose is not an EPA final approved value

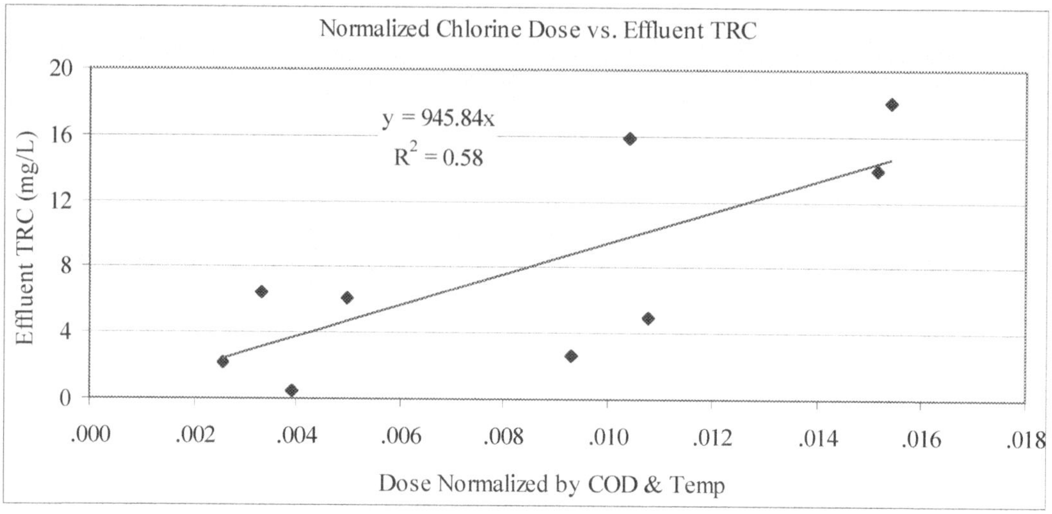

Figure 4-6. Normalized Chlorine Dose and Pre-Dechlorination Effluent TRC.

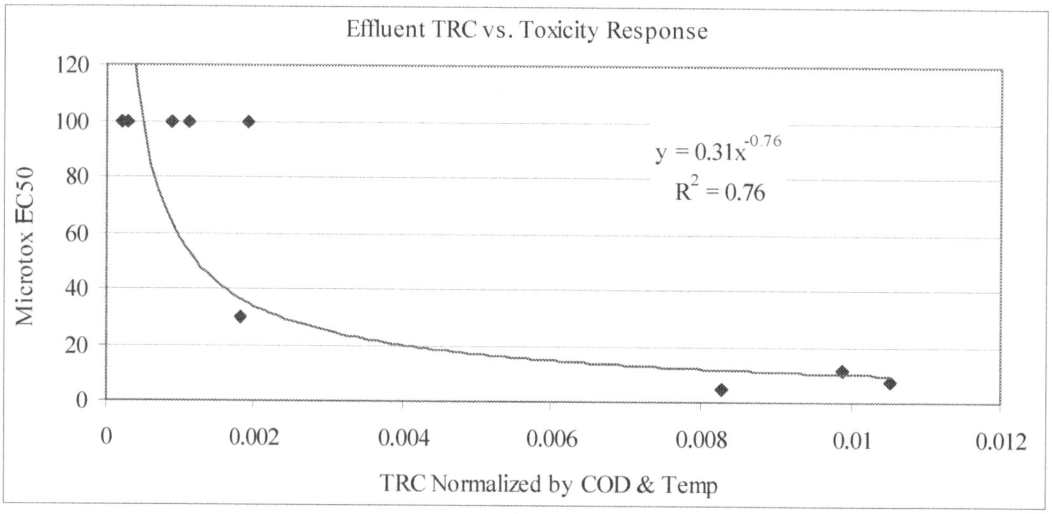

Figure 4-7. Normalized Post-Dechlorination Effluent TRC and Effluent Toxicity.

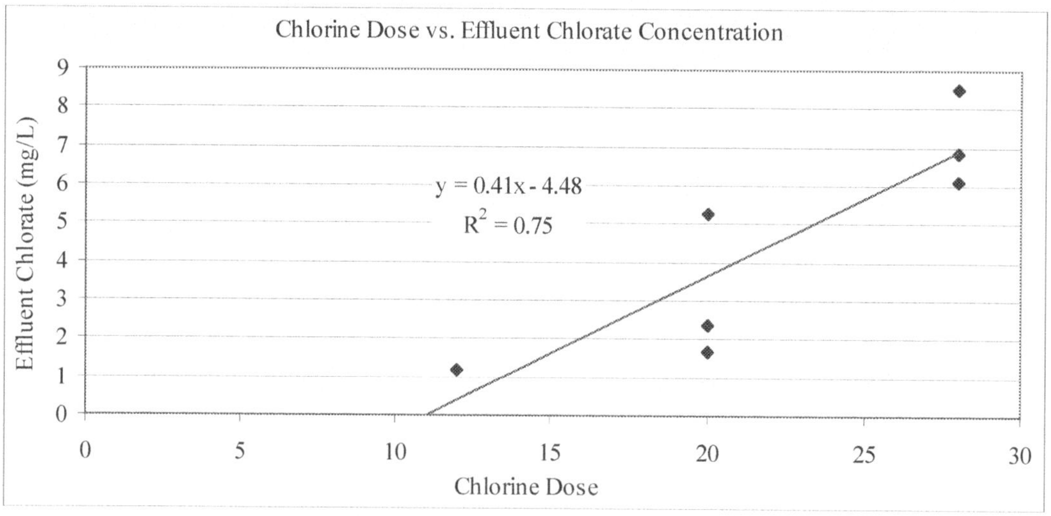

Figure 4-8. Chlorine Dose and Effluent Chlorate.

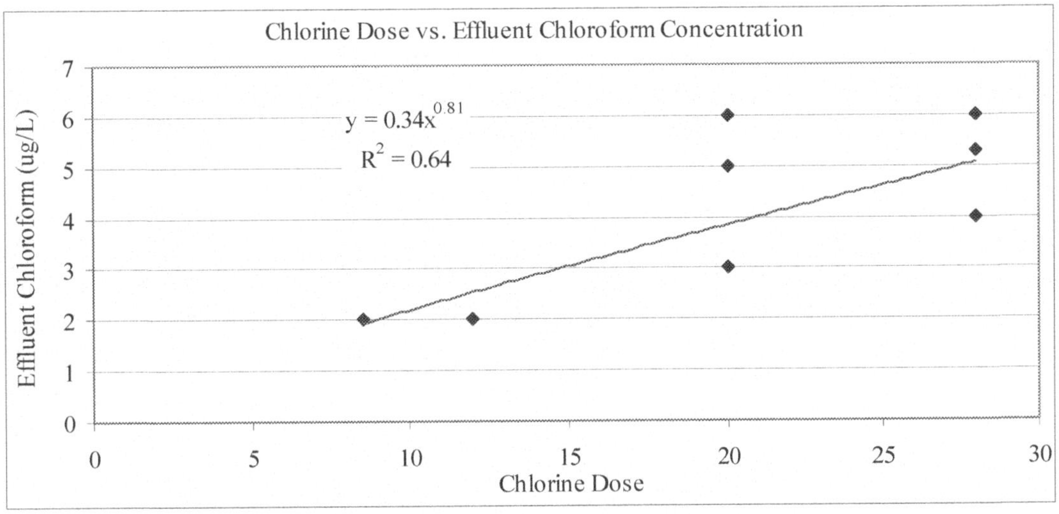

Figure 4-9. Chlorine Dose and Effluent Chloroform.

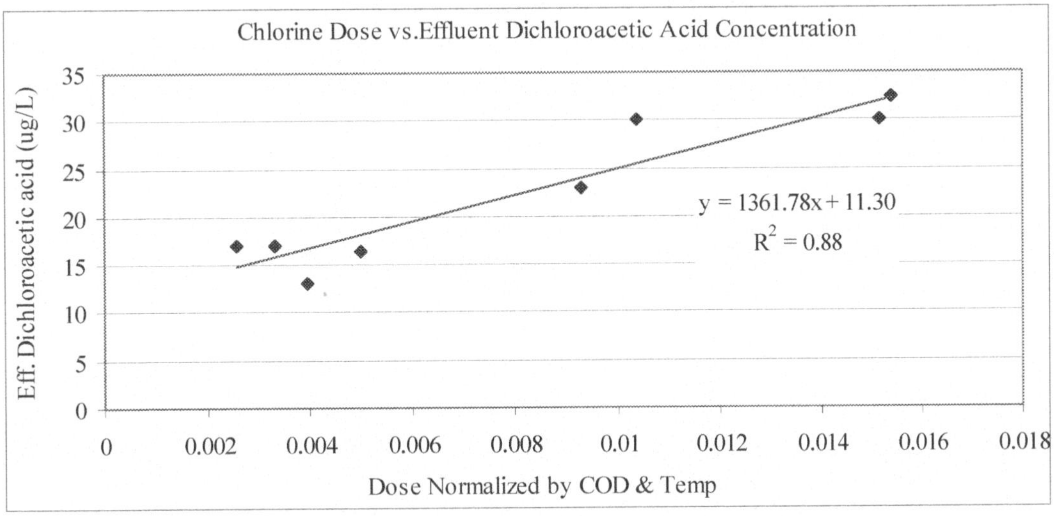

Figure 4-10. Normalized Chlorine Dose and Effluent Dichloroacetic Acid.

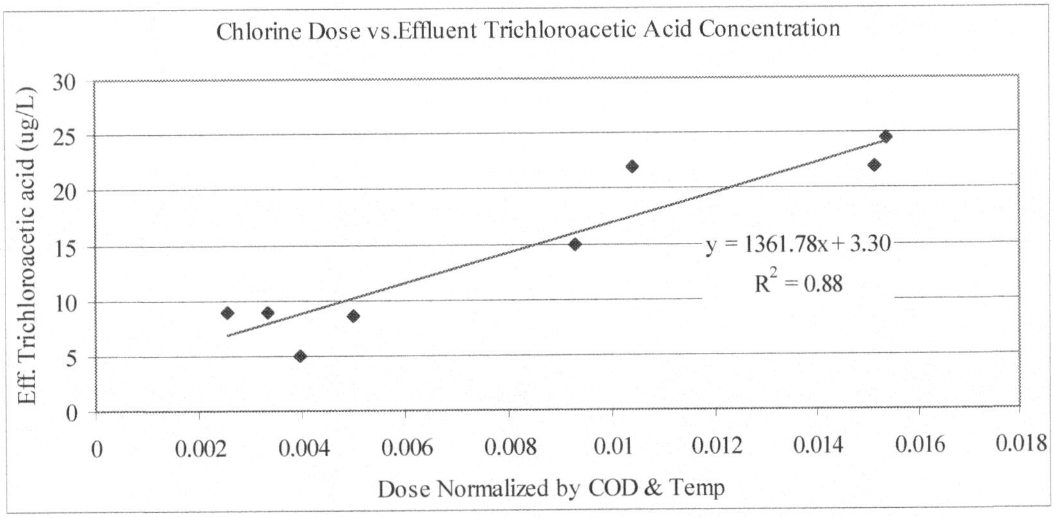

Figure 4-11. Normalized Chlorine Dose and Effluent Trichloroacetic Acid.

4.3.4 Chlorine Dioxide

Disinfection with chlorine dioxide was tested on eleven wet weather flow samples (event eight was run under three different chlorine dioxide doses). The applied dose concentrations ranged from 1 to 15 mg/L. The chlorine dioxide was applied with a high-RPM mixer and 5 min of contact time was allowed. The ranges of percent reduction for the three types of bacteria are summarized in Table 4-9. In terms of percent reduction, chlorine dioxide was most effective in inactivating fecal coliform and least effective in inactivating *Enterococcus* spp.

Table 4-9. Bacterial Data Summary for Chlorine Dioxide Disinfection.

Bacterial Type	Initial Count Range (CFU/100 mL)	% Reduction Range	Avg. Log Reduction
Fecal coliform	2.8E5 to 2.4E7	80.451 to 99.999	3.0
E. coli	6.0E4 to 1.6E7	81.061 to 99.999	2.7
Enterococcus spp.	4.7E3 to 7.2E6	84.211 to 99.967	2.4

4.3.4.1 Dose-Response Relationships

Scatter plots were developed for chlorine dioxide dose versus log of effluent bacterial concentration and log bacterial reduction for each of the three bacterial groups. Each of the dose-response curves could be improved by normalizing the dose by influent water quality parameters, including COD, BOD, iron, TSS, TOC, and temperature. Similar to the chlorination disinfection data, applied dose normalized by COD and temperature was used to develop dose-response curves for deriving the target doses for achieving the desired bacterial concentrations.

Dose-response curves normalized by COD and temperature are shown for the three indicator bacterial organisms in Figure 4-12. The dose-response relationships are strong for each of the three bacterial types (R^2 values of 0.72 or greater). Each of the dose-response curves for chlorine dioxide resulted in higher R^2 values than the curves developed for chlorination. The strongest relationship was usually observed using a power regression equation. Similar to the chlorination disinfection data, the correlations become more significant as the indicator organism becomes more specific. Both *E. coli* and *Enterococcus* spp. show significant correlation with dose normalized by COD and temperature.

Assuming the same COD and temperature as the typical wet weather flow at METRO, these plots indicate that it will require approximately 25% less chlorine dioxide to achieve a 1,000 CFU/100 mL for *Enterococcus* spp. than it would take to achieve a 1,000 CFU/100 mL for fecal coliform, which is likely due to the influent challenge concentration being higher for the fecal coliform group. However, to achieve the maximum value criteria for each indicator organism (4,274 CFU/100 mL for fecal coliform, 2,507 CFU/100 mL for *E. coli*, and 656 CFU/100 mL for *Enterococcus* spp.) approximately 36% more chlorine dioxide is required to inactivate *Enterococcus* spp. as compared to fecal coliform. The chlorine dioxide dose necessary to achieve bacterial concentrations of 1,000 CFU/100 mL and the maximum value criteria are summarized in Table 4-10. Again, the values are calculated from the dose-response curves normalized for a COD of 200mg/L and temperature of 20°C.

Table 4-10. Chlorine Dioxide Dose Required to Meet Bacterial Criteria.

Parameter	ClO_2 Dose to Achieve Maximum Value Criteria	Maximum Value Criteria	ClO_2 Dose to Achieve 1,000 CFU/100 mL
Fecal coliform	6.3 mg/L	4,274 CFU/100 mL	9.6 mg/L
E. coli	6.6 mg/L	2,507 CFU/100 mL	8.7 mg/L
Enterococcus spp.	8.6 mg/L	656 CFU/100 mL	7.2 mg/L

Note: The above table uses the correlations based upon dose normalized by COD and temperature.

Lower chlorine dioxide doses were required to achieve the same bacterial reductions compared to the chlorination disinfection process. In general, chlorine dioxide in water is 10 times more soluble, a broader bactericide, and an effective viricide than chlorine. It does not react with ammonia nitrogen or oxidizable material and is relatively unaffected in the pH range of 6 to 10. In many cases, it is considered a more powerful oxidant than chlorine because the chlorine atom undergoes five valence changes in the process of oxidation to the chloride ion, whereas one mole of hypochlorous acid receives only two electrons (White, 1999).

$$ClO_2 + 5e^- + 2H_2O = Cl- \ + 4OH^-$$
$$HOCl + H+ + 2e^- \ = \ Cl- \ + \ H2O$$

This indicates that chlorine dioxide theoretically has about 2.5 times the oxidizing power of chlorine. However, the oxidizing capacity of chlorine dioxide is not all used in water disinfection because the majority of the chemical reactions result in chlorine dioxide being reduced to chlorite.

$$ClO_2 + 1e^- \quad = ClO_2^-$$

Relationships between fecal coliform, *E. coli*, and *Enterococcus* spp. in the disinfected effluent are shown in Figure 4-13. The chlorine dioxide disinfected effluent shows strong correlations between all three indicator bacterial organisms. These relationships indicate that fecal coliform data from a chlorine dioxide treated effluent can be transformed into *E. coli* or *Enterococcus* spp. with some level of certainty. Additionally, comparing the slopes of the lines from the influent charts (Figure 4-2) to the effluent charts (Figure 4-13) provides an indication of the effectiveness of chlorine dioxide to inactivate the different bacterial concentrations. The slopes of the lines for the effluent charts are similar to the slopes of the lines for the influent charts, which indicate that chlorine dioxide was equally effective at inactivating all the bacterial groups. This outcome is different from the outcome to achieve the 1,000 CFU/100 mL criteria, which indicates that chlorine dioxide was most effective at inactivating *Enterococcus* spp. The difference is likely due to the variability in the data and the relatively tight range of doses, 7.2 mg/L –9.6 mg/L (Table 4-10).

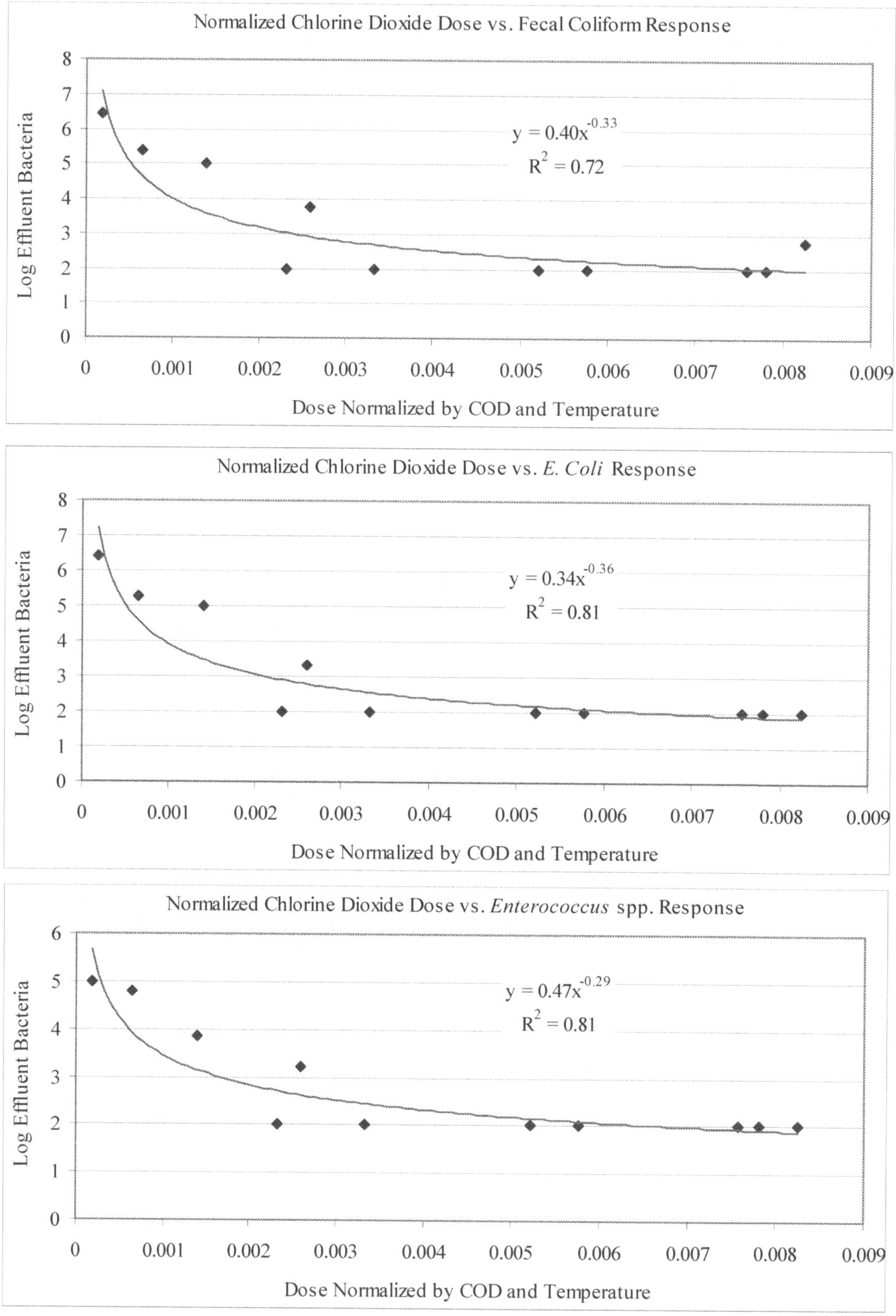

Figure 4-12. Chlorine Dioxide Dose-Response Curves Normalized by COD/Temperature.

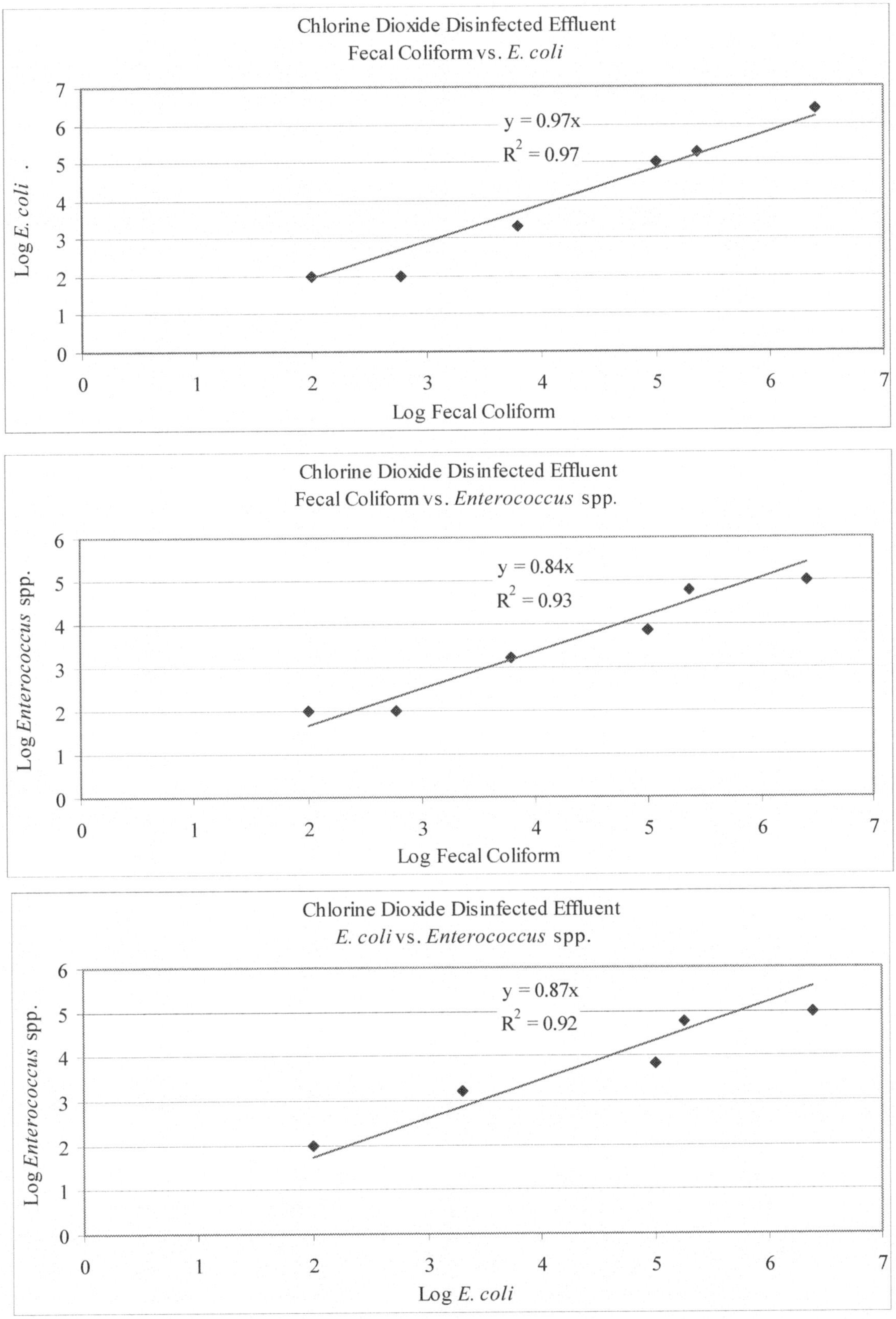

Figure 4-13. Comparison of the Bacterial in the Chlorine Dioxide Effluent.

4.3.4.2 Disinfection Residuals and By-products

The concentrations of disinfection residuals and DBPs were measured in the influent and effluent of the chlorine dioxide pilot for each of the sampling events. The effluent sampling results are presented in Table 4-11. Ferrous sulfate was used to reduce the residual chlorite levels in events five through nine; however only during events five, eight, and nine was ferrous sulfate added in significant quantities to reduce chlorites prior to their measurement. It appears that during these events the ferrous sulfate also reduced the chlorine dioxide that otherwise would have been left as residual. Similar to chlorination/dechlorination, aquatic toxicity was measured after the addition of the reducing agent (i.e., ferrous sulfate).

Table 4-11. Chlorine Dioxide Effluent Disinfection Residual and DBP Data.

Event	Event 1	Event 2	Event 3	Event 4	Event 5	Event 6	Event 7	Event 8	Event 9
Dose (mg/L)	1	5	10	15	5	10	15	5	5
Chlorate (mg/l)	1.7	1.7	6.1	7	2	4.4	13.4	2.4	2.52
Chlorite (mg/l)	11.8	11.8	25.9	32.7	<0.5	26.8	27	<0.025	<0.6
TRC - post-ferrous sulfate (mg/l)	0.41	0.41	1.41	3.46	0.04	<0.03	3.76	<0.03	0.39
Residual Chlorine Dioxide (mg/l)	0.06	0.06	0.82	4.58	<0.03	1.73	7.12	<0.03	<0.03
Chloroform (ug/l)	2	2	1.7	2	2	1	3	3	3
Bromodichloromethane (ug/l)	<1	<1	<0.5	<1	<1	<1	<1	<1	<1
Dibromochloromethane (ug/l)	<1	<1	<0.5	<1	<1	<1	<1	<1	<1
Bromoform (ug/l)	<1	<1	<0.5	<1	<1	<1	<1	<1	<1
Dibromoacetic acid (ug/l)	<1	1.4	<1	<1	<1	6.3	2.2	<1	<1
Dichloroacetic acid (ug/l)	1.3	69	100	86	12	49	61	13	48
Monobromoacetic acid (ug/l)	<1	<1	<1	<1	<1	6.5	2.5	<1	16
Monochloroacetic acid (ug/l)	<2	28	17	16	2	12	15	2.4	5.5
Trichloroacetic acid (ug/l)	2.8	6	32	13	14	12	8.1	5.9	11
Acute Toxicity - Fathead Minnow	<100	<<100	31.9	>10 <100	<71	>14 <71	ND	ND	ND
Acute Toxicity - *C. dubia*	<100	<<100	3.12	<0.25	<71	<7	<3.5	>14 <100	>100
Microtox™	37	50-100	13	10.2	>100	32.2	12.2	>100	~100

Similar to the dose-response curves, disinfection residuals and DBPs were plotted against dose and dose normalized by COD and temperature. The curve that provided the best correlation was used to quantify the level of residuals and DBPs found in the effluent.

Residual chlorine dioxide is produced at increasing rates to the influent dose as shown in Figure 4-14. The residual chlorine dioxide data presented in Figure 4-14 are from events one through four and six and seven when either no ferrous sulfate or an insignificant amount of ferrous sulfate was added. These data were chosen so that the relationship between dose and residual without the effect of a reducing agent could be portrayed. When using a chlorine dioxide dose of 6.6 mg/L (the required dose to achieve the maximum *E. coli* criterion of 2,507 CFU/100 mL for a typical wet weather flow at METRO — Table 4-10), the residual chlorine dioxide is estimated from Figure 4-14 to be approximately 0.14 mg/L.

Chlorine dioxide residual concentrations ranged from less than 0.1 to 7.0 for samples in which ferrous sulfate was not added or added in insignificant quantities. The maximum residual disinfection level goal (MRDLG) for residual chlorine dioxide, established by the U.S. EPA Microbial and Disinfection By-product Rule for drinking water, is 0.8 mg/L. Based on a dose of 6.6 mg/L and a resulting residual chlorine dioxide of 0.14 mg/L this criterion would not be exceeded. Nevertheless, it appears that during events five, eight and nine the ferrous sulfate successfully reduced the chlorine dioxide that otherwise would have been left as residual.

Chlorine dioxide residuals are compared to the Microtox EC_{50} and shown in Figure 4-15. Figure 4-15 indicates that any significant chlorine dioxide residual will cause increased aquatic toxicity. However, it is important to note that in the case of chlorine dioxide the cause of toxicity is likely a mixture of residual chlorine dioxide, residual chlorine, and chlorite. Because of this mixture of chemicals, only antidotal evidence can be offered as to the likelihood of one chemical resulting in more toxicity than the other. For example, in events one and two significant toxicity was observed and the residual chlorine dioxide was only 0.06 mg/L. During these events the residual chlorine was 0.41 mg/L and the chlorite concentrations were 12 and 10.4, respectively. Based on literature findings, it is unlikely that a chlorine dioxide residual of 0.06 mg/L and a chlorine residual of 0.41 mg/L would cause this level of toxicity. Therefore chlorite was likely the largest contributor to toxicity; chlorite is discussed below. Nevertheless, sensitive receiving waters may require residual chlorine dioxide removal to minimize aquatic toxicity. This determination should be made by examining the dynamics of the discharge and receiving water and should consider allowable mixing zones.

The primary DBPs of chlorine dioxide disinfection are chlorate (CLO_3^-) and chlorite CLO_2^-) ions. The production of these two constituents compared to chlorine dioxide dose is presented in Figure 4-16 and Figure 4-17 for chlorate and chlorite, respectively. These charts only include data from samples in which ferrous sulfate was not added or added in insignificant quantity.

The amount of chlorate produced ranged from 1.7 to 13 mg/L and correlated well with chlorine dioxide dose. No MCL or MCLG is regulated by the EPA for chlorate under the Microbial and Disinfection By-product Rule or the SDWA. A chlorine dioxide dose of 6.6 mg/L (the required dose to achieve the maximum *E. coli* criterion of 2,507 CFU/100 mL for a typical wet weather flow at METRO — Table 4-10) would result in a chlorate increase of approximately 3 mg/L.

The chlorite concentrations are approximately five times the chlorate concentration in six of the nine events where ferrous sulfate was not added or added in insignificant quantity. Additionally, effluent chlorite concentrations were found to be linearly proportional to the chlorine dioxide dose. Chlorite concentrations of approximately 10 mg/L were observed at even the lowest applied dose, which indicates that chlorite was present in the chlorine dioxide product after the generation process (i.e., acidification of sodium chlorite). Additionally, based on the residual chlorine concentrations in the effluent, it is also clear that chlorine was another undesirable by-product of the generation method.

The MCLG for chlorite in drinking water, established by the U.S. EPA Microbial and Disinfection By-product Rule, is 0.8 mg/L. A chlorine dioxide dose of 6.6 mg/L (the required dose to achieve the maximum *E. coli* criterion of 2,507 CFU/100 mL for a typical wet weather flow at METRO — Table 4-10) would result in a chlorite increase of approximately 18 mg/L before the addition of ferrous sulfate. The chlorite concentration (18 mg/L) in the chlorine dioxide effluent exceeded the applied dose (6.6 mg/L) because there was residual chlorite in the chlorine dioxide product as a result of the generation method employed for this demonstration. Figure 4-18 indicates that any significant chlorite will cause increased aquatic toxicity. Based on chlorite concentration of 18 mg/L, the Microtox EC_{50} would reduce from greater than 100% to approximately 46%, as depicted in Figure 4-18. As discussed under chlorine dioxide residual, it is important to note that a mixture of residual chlorine dioxide and chlorite was likely the cause of the toxicity. However, chlorite may have been a more lethal component of the mixture.

As discussed in Section 4.2.1, chlorine dioxide was generated by acidifying sodium chlorite, which was the most appropriate method for this small-size demonstration. Other methods of chlorine dioxide generation, such as methods that use chlorine gas to oxidize sodium chlorite, do not leave residual chlorite in the product. Such a process was used during a disinfection demonstration performed for New York City Department of Environment Protection (Camp Dresser & McKee and Moffa & Associates, 1997). In the New York City demonstration, when applied doses ranged from 5 to 6 mg/L, the resulting average chlorite concentration was 1 mg/L, or almost 20 times less than measured during this demonstration. When considering only marginal receiving water dilution (Section 5.2.3) the chlorite concentrations would be less than that of U.S. EPA Microbial and Disinfection By-product Rule, 0.8 mg/L.

THMs were not produced by chlorine dioxide disinfection.

On average, the total HAA effluent concentration was approximately six times the influent concentrations. The three specific HAAs produced were mono-, di-, and trichloroacetic acids. The mono-, di-, and trichloroacetic acids were plotted against the dose as shown in Figure 4-19, 4-20 and 4-21, respectively. The correlations show a general trend that indicates these substances are produced as a power function with respect to the applied dose. The total HAA effluent concentrations ranged between 8.1 and 151 µg/L. The SDWA MCL for total HAA is 60 µg/L. This significant increase in HAAs is counter to the literature (Werdehoff and Singer, 1987), which documents that chlorine dioxide does not chlorinate organics and therefore does not produce THMs and HAAs. It is possible that chlorine, as evidenced by the TRC concentrations in the effluent, is reacting with organics to produce HAAs. Further testing is suggested using chlorine dioxide generated from other methods to examine the production of HAAs produced in wet weather flow disinfection.

A summary of residual chlorine dioxide and DBPs is presented in Table 4-12. The values in this table were derived for typical wet weather flow quality at METRO, 200 mg/L COD and 20°C, and an applied dose of 6.6 mg/L as required to achieve the maximum value *E. coli* criterion. The parameters, with the exception of THMs and HAAs, reflect concentrations before the addition of ferrous sulfate.

Table 4-12. Chlorine Dioxide Residual and DBP Concentrations for Typical Wet Weather Flow at METRO.

Parameter	Value
Dose	6.6 mg/L
Residual Chlorine Dioxide	0.14 mg/L
Chlorate	3.3 mg/L
Chlorite	18 mg/L
Microtox (EC_{50}%)	46%[1]
HAA-Monochloroacetic acid	6 µg/L
HAA-Dichloroacetic acid	40 µg/L
HAA-Trichloroacetic acid	12 µg/L

1. Microtox EC_{50} based on chlorite as presented in Figure 4-17 and 4-18.

Note: The dose in the above table is the dose required to obtain an *E. coli* concentration of 2,507 CFU/100 mL. The chlorite concentration is before the addition of ferrous sulfate and the resulting Microtox is based on 18 mg/L chlorite. It is important to note that the aquatic toxicity is likely the result of chlorite as well as residual chlorine dioxide and chlorine. The chlorite concentration exceeded the applied dose because there was residual chlorite in the chlorine dioxide product as a result of the generation method employed for this demonstration.

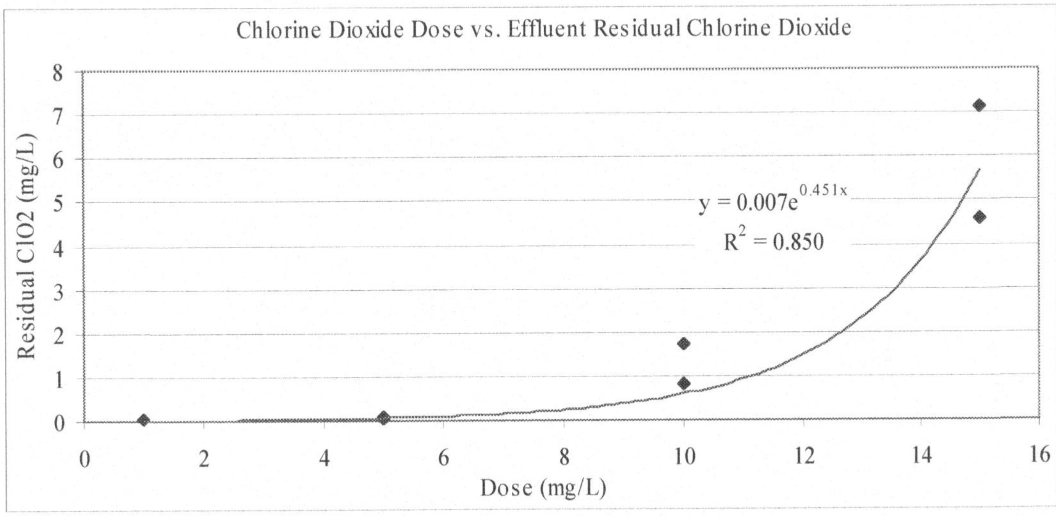

Figure 4-14. Chlorine Dioxide Dose and Pre-Reducing Agent Effluent Residual Chlorine Dioxide.

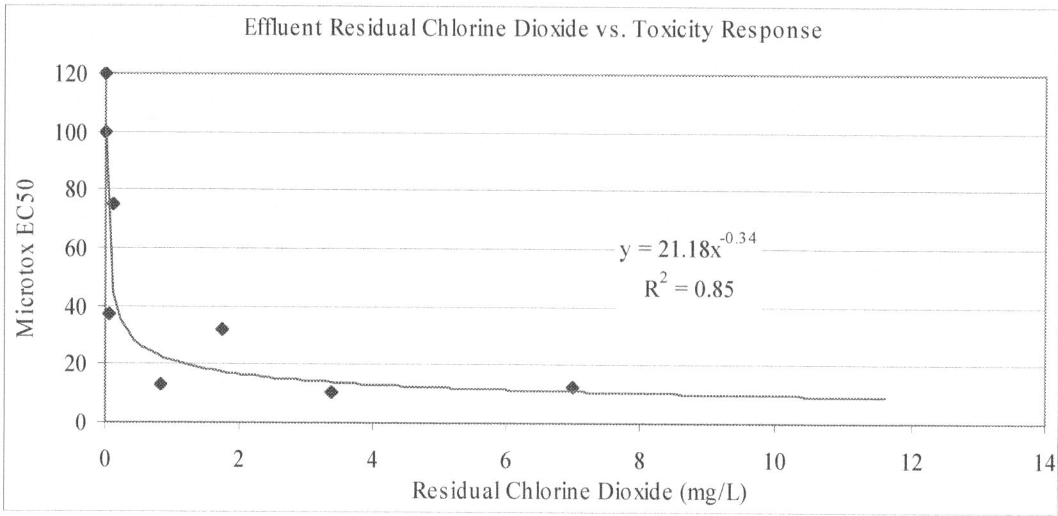

Figure 4-15. Post-Reducing Agent Effluent Chlorine Dioxide Residual and Toxicity.

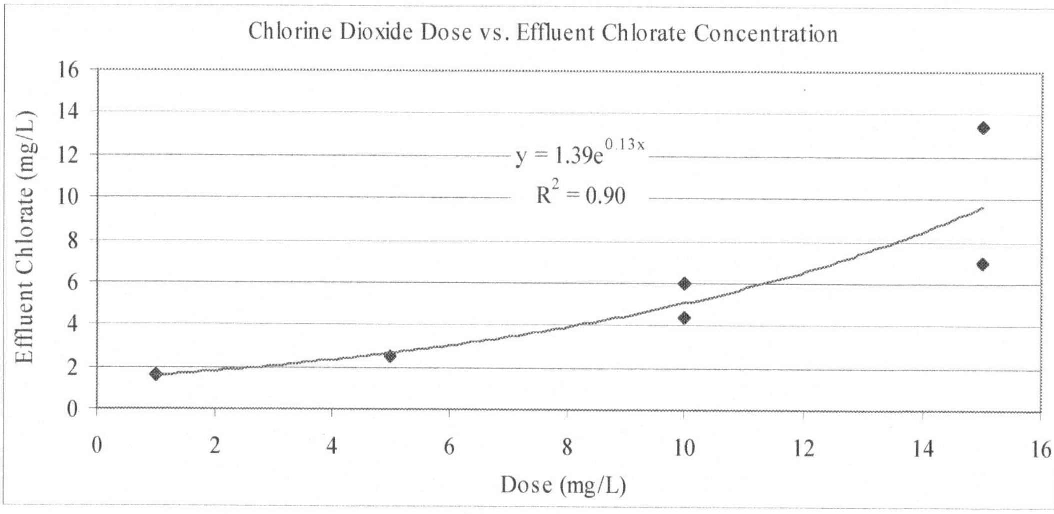

Figure 4-16. Chlorine Dioxide Dose and Effluent Chlorate.

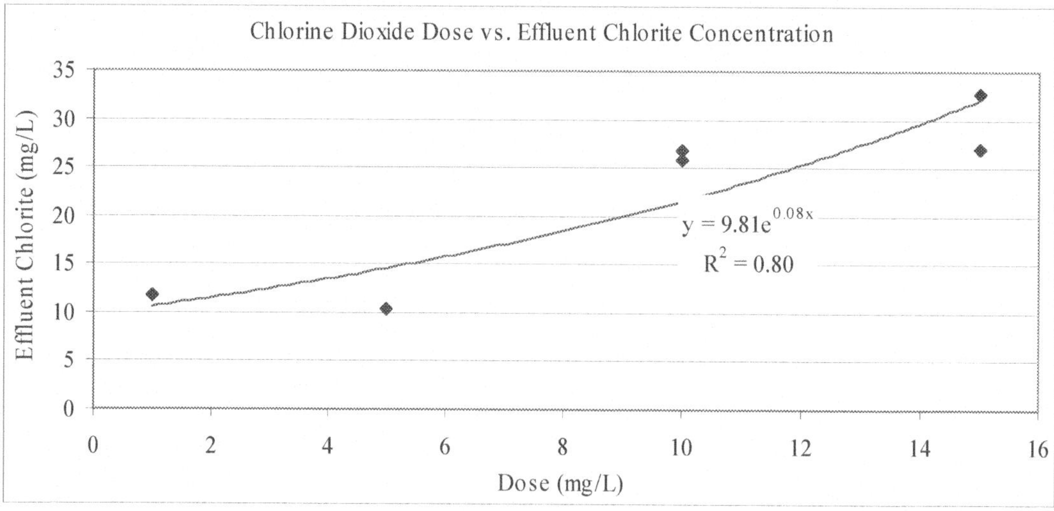

Figure 4-17. Chlorine Dioxide Dose and Pre-Reducing Agent Effluent Chlorite.

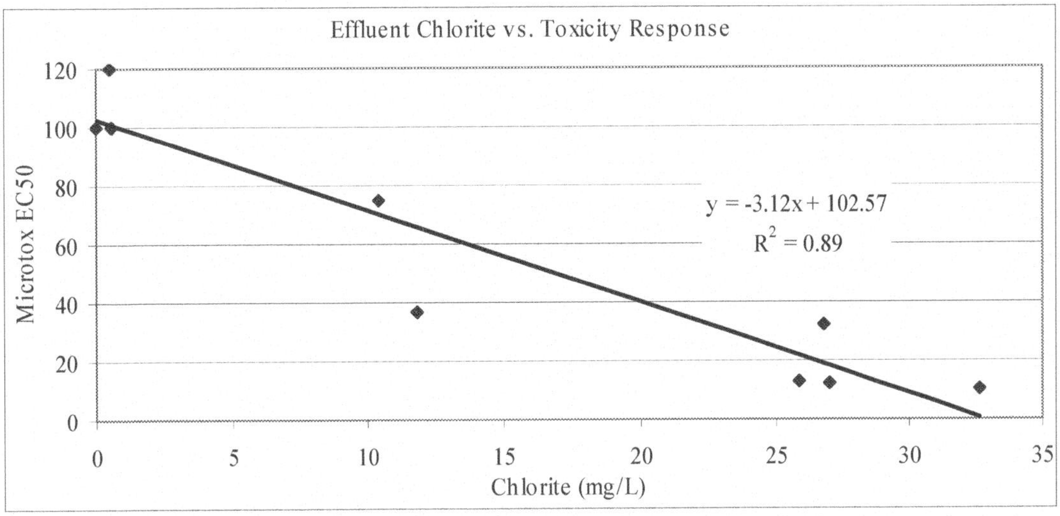

Figure 4-18. Post-Reducing Agent Effluent Chlorite and Toxicity.

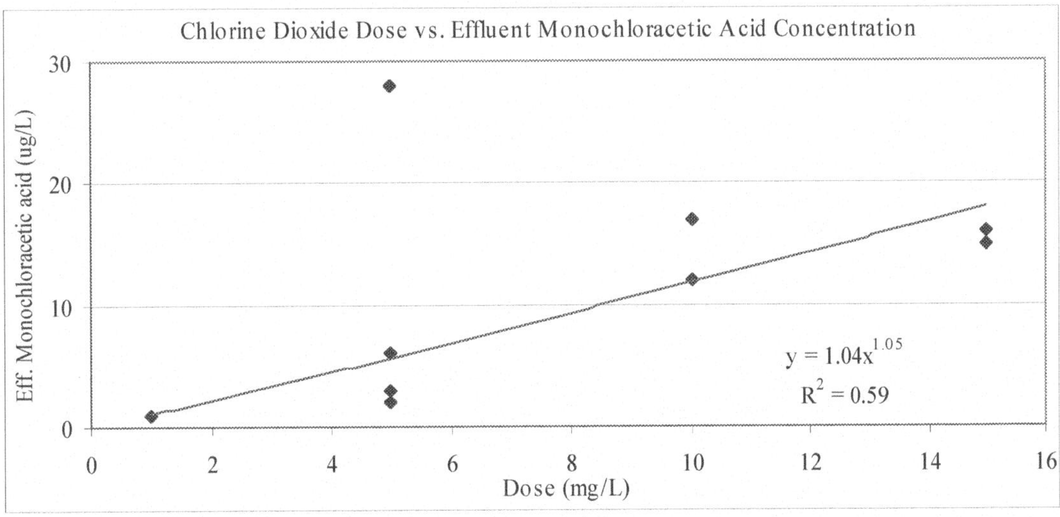

Figure 4-19. Chlorine Dioxide Dose and Effluent Monochloroacetic Acid.

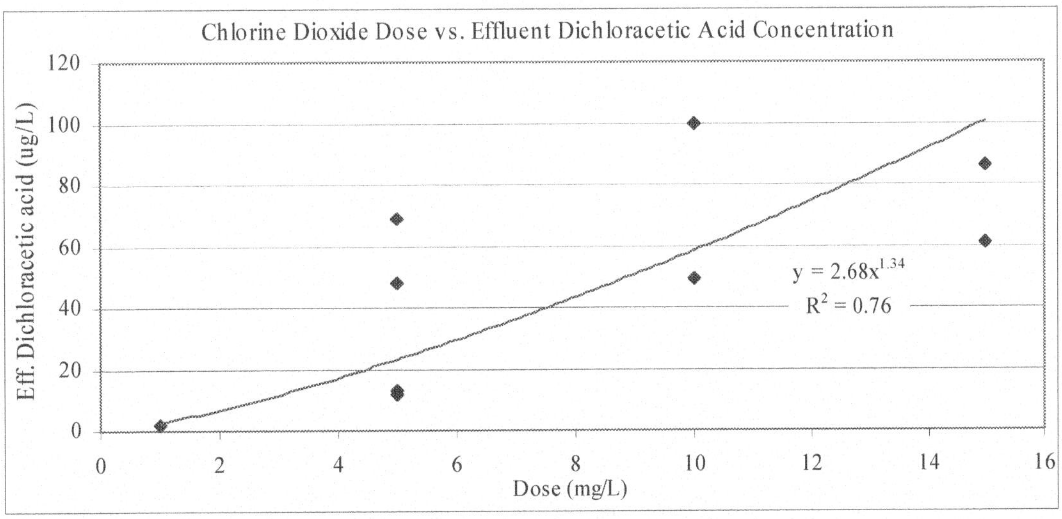

Figure 4-20. Chlorine Dioxide Dose and Effluent Dichloroacetic Acid.

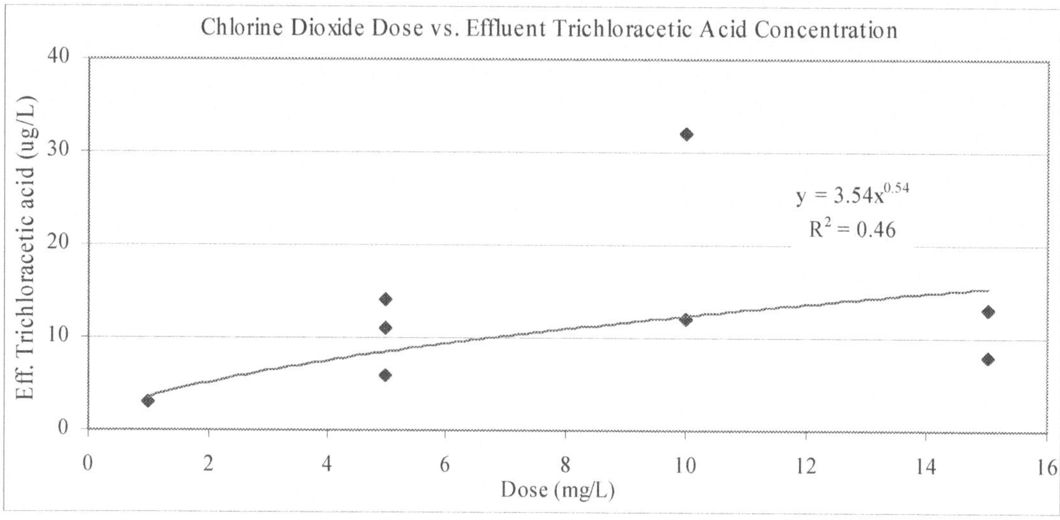

Figure 4-21. Chlorine Dioxide Dose and Effluent Trichloroacetic Acid.

4.3.4.3 Toxcity

Measured Aquatic Toxicity

Chlorine dioxide-treated effluent was toxic to each species in each event in which they were tested, with the exception of *C. dubia* for event nine and the Microtox bacteria for events five, eight, and nine. An LC_{50} with minnows was obtained in event three (32%). In all other instances during the first six events, LC_{50}s with the minnows were estimated from screening test results ranging from > 10% to < 100%, with the most probable range of LC_{50}s being 20% to 40%. Minnows were excluded from testing in the last three events. *C. dubia* appeared to be significantly more sensitive to chlorine dioxide treated samples than minnows with LC_{50}s ranging from < 0.25% to < 100% for events one through eight, with the most probable range of LC_{50}s being 5% to 10%. *C. dubia* did not show signs of toxicity during event nine. Microtox appeared to be less sensitive than either fish or *C. dubia.* During events one through four and events six and seven, Microtox EC_{50}s ranged from 10% to $\leq$ 100% at levels slightly higher (less sensitive) to that observed with minnows and *C. dubia*. Microtox EC_{50}s during events five, eight, and nine were approximately 100%.

It is conjectured that the aquatic toxicity associated with chlorine dioxide effluent is the result of the relatively high chlorite concentrations present as a residual of the chlorine dioxide generation method as discussed in Section 4.4.2. Ferrous sulfate was used to reduce the chlorine dioxide and chlorite residuals in the effluent. In three of the nine events (five, eight, and nine), ferrous sulfate was added in significant quantities to reduce chlorites in the effluent. During these events, toxicity as measured by the Microtox bacterial assay was reduced.

Literature Review Toxicity Findings

A summary of literature review toxicity data for chlorine dioxide and DBPs is presented in Table 4-13. Table 4-13 only shows residuals and DBPs that increased in concentration as a result of the chlorine dioxide process. Additional information on these chemicals and other chemicals can be found in Chapter 8.0. Table 4-13 presents acute and chronic toxicity values for fresh and marine water animals; the circled values are the lowest or most conservative values found. For mammalian toxicity, reference doses for chlorine dioxide, chlorite, monochloroacetic acid, and dichloroacetic acid were found and a cancer slope factor for dichloroacetic acid was found. Chlorine dioxide and DBPs may cause cancer in mammalian species, but the current database is inconclusive. Some of the organs affected by these chemicals include bladder, liver, and kidney.

Table 4-13. Toxicity Literature Review Findings for Chlorine Dioxide.

		Acute Toxicity Value			Chronic Toxicity Value			Mammalian Toxicity			
Chemical	**CAS NO**	**Species**	**Value (μg/L)**	**Endpoint**	**Species**	**Value (μg/L)**	**Endpoint**	**Carcinogen**	**Possible Organs Affected**	**RfD (mg/Kg-day)**	**CSF $(mg/Kg\text{-}day)^{-1}$**
Chlorine dioxide	10049-04-4	FW Fish: FW Invert: SW Fish: SW Invert:	6400000 (350) ND 500000	96-hr LC50 96-hr LC50 ND 48-hr LC50		ND		ND	ND	$3x10^{-2}$	ND
Chlorate ion	14866-68-3	ND			ND			ND	ND	ND	ND
Chlorite ion	7758-19-2	FW Fish: FW Invert: SW Fish: SW Invert:	100000 (10) 75000 580	96-hr LC50 48-hr EC50 96-hr LC50 96-hr LC50	FW Fish: FW Invert: SW Fish: SW Invert:	50600 ND ND ND	312-hr LC50 ND ND ND	ND	ND	$3x10^{-2}$	ND
Monochloroacteic acid (metabolite of DCA)	79-11-8	FW Fish: FW Invert: SW Fish: SW Invert:	ND (77000) ND ND	ND 48-hr EC50 ND ND	FW Fish: FW Invert: SW Fish: SW Invert:	ND 32000 ND ND	ND 504-hr NOEC ND ND	ND	liver, testicle, heart, vessels, kidney	0.002 (1)	ND
Dichloroacetic acid (DCA)	79-43-6	FW Fish: FW Invert: SW Fish: SW Invert:	ND ND ND (23000)	ND ND ND 96-hr LC50		ND		Probable	liver, testicle, heart, vessels, kidney	$4x10^{-3}$	$5x10^{-2}$
Trichloroacetic acid	76-03-9	FW Fish: FW Invert: SW Fish: SW Invert:	1050000 (1200) 9300000 4800000	96-hr LC50 24-hr LC50 96-hr LC50 96-hr LC50		ND		Possible	liver, testicle, heart, vessels, kidney	ND	ND

Circled value is the lowest available ecotoxicity value and is therefore used as the screening toxicity value for the risk assessment.
ND = No Data or Insufficient Data
RfD = Reference Dose
CSF = Cancer Slope Factor
(1) This reference dose is not an EPA final approved value

4.3.5 Ozonation

Disinfection with ozone was tested on twelve wet weather flow samples (no samples were taken during event one and three different disinfection doses were applied and sampled during events four and nine). Ozone gas was introduced into the water with an eductor and 3-5 minutes of contact time was provided. The dosing concentration ranged from 4.6 mg/L to 35.1 mg/L. The ranges of percent reduction for the three types of bacteria are summarized in Table 4-14. In terms of percent reduction, ozonation was equally effective in inactivating all three indicator bacterial organisms.

Table 4-14. Bacterial Data Summary for Ozonation Disinfection.

Bacterial Type	Initial Count Range (CFU/100 mL)	% Reduction Range	Avg. Log Reduction
Fecal coliform	1.0E5 to 2.7E7	85.185 to 99.833	2.0
E. coli	3.0E4 to 1.2E7	42.857 to 99.800	2.1
Enterococcus spp.	5.1E3 to 4.3E5	-9.3 to 99.789	2.1

Compared to the other chemical disinfectants, ozonation had the highest average effluent bacteria concentrations and the lowest average log reductions. The difference in effectiveness is especially apparent for the fecal coliform and *Enterococcus* spp., which have average log reduction of approximately two as compared to the other disinfectants that had log reductions of approximately three. This is contrary to the literature regarding ozonation (Morris, 1975). Research by J.C. Morris illustrates that ozone is a more powerful germicide against all classes of organisms by a factor of 10–100 times. This contrast may be the result of difficulties measuring ozone residuals and hence dose or the fact that ozone was readily consumed by constituents in the water.

4.3.5.1 Dose-Response Relationships

Scatter plots were developed for ozone dose versus log of effluent concentration and log reduction for each of the three bacterial groups. Linear, logarithmic, and power regressions were tested against each data set, and the strongest relationship was used for calculating the coefficient of determination shown on each plot. The strongest relationship was usually observed using a power regression equation.

In order to properly compare the difference of ozonation to the other disinfectants, the "Event Five" data point was removed from the correlation plot. Field records noted difficulty with measuring wastewater flow rate and off-gas concentration, and it was documented that the actual dose was likely greater than the reported value. In light of these documented difficulties, it is appropriate to consider these data as outliers and therefore remove them from the analysis. The design practitioner may also consider removing event six in every case. This would result in a conservative approach and provide a factor of safety for the facility design. However, the data from "Event Six" were not removed from this analysis.

Each of the dose-response curves could be improved by normalizing the applied dose with a variety of influent water quality parameters including COD, BOD, iron, TSS, TOC, and temperature. Similar to the chlorination disinfection process, applied dose normalized by COD and temperature was used to develop dose-response curves for deriving the appropriate doses for achieving the desired bacterial concentrations. Also similar to the chlorination disinfection process, applied dose normalized by TSS and temperature was used to develop an algorithm that can be used for process control with instrumentation such as TSS or turbidity meters (the development of this algorithm is discussed in Chapter 5.0).

Dose-response curves normalized by COD and temperature are shown for the three indicator bacterial organisms in Figure 4-22. The poorer correlations for the ozonation disinfection process may also be related to the difficulty of measuring residual due to its low solubility. This also makes ozonation difficult to control.

For the typical wet weather flow at METRO, these plots indicate that it will require approximately 50% less ozone to achieve a 1,000 CFU/100 mL for *Enterococcus* spp. as compared to achieving 1,000 CFU/100 mL for fecal coliform, which is likely due to the influent challenge concentration being higher for the fecal coliform group. To achieve the maximum value criteria for each indicator organism (4,274 CFU/100 mL for fecal coliform, 2,507 CFU/100 mL for *E. coli*, and 656 CFU/100 mL for *Enterococcus* spp.) it requires approximately 22% less ozone to inactivate *Enterococcus* spp. as compared to fecal coliform. This is counter to the trends observed in chlorination, chlorine dioxide and UV, and may be the result of difficulties measuring ozone residuals and hence dose. The ozone dose necessary to achieve bacterial concentrations of 1,000 CFU/100 mL and the maximum value criteria are summarized in Table 4-15. Again, the values are calculated from the dose-response relationship normalized for a COD of 200 mg/L and temperature of 20°C.

Table 4-15. Ozone Doses Required to Meet Bacterial Criteria.

Parameter	O_3 Dose to Achieve Maximum Value Criteria	Maximum Value Criteria	O_3 Dose to Achieve 1,000 CFU/100 mL
Fecal coliform	25 mg/L	4,274 CFU/100 mL	35 mg/L
E. coli	23 mg/L	2,507 CFU/100 mL	28 mg/L
Enterococcus spp.	20 mg/L	656 CFU/100 mL	17 mg/L

Note: The above table uses the correlations based upon dose normalized by COD and temperature.

Relationships between fecal coliform, *E. coli*, and *Enterococcus* spp. in the disinfected effluent are shown in Figure 4-23. Similar to the influent data (Section 4.3.2), these relationships indicate that fecal coliform data from a ozonated effluent can be transformed into *E. coli* with some level of certainty (excluding the event five outlier). Additionally, comparing the slopes of the lines from the influent charts (Figure 4-2) to the effluent charts (Figure 4-23) provides an indication of the effectiveness of ozone. Based on changes to the slope of the lines, ozonation was most effective at inactivating *Enterococcus* spp., which was likely because fecal coliform and *E. coli* had the highest challenge concentrations.

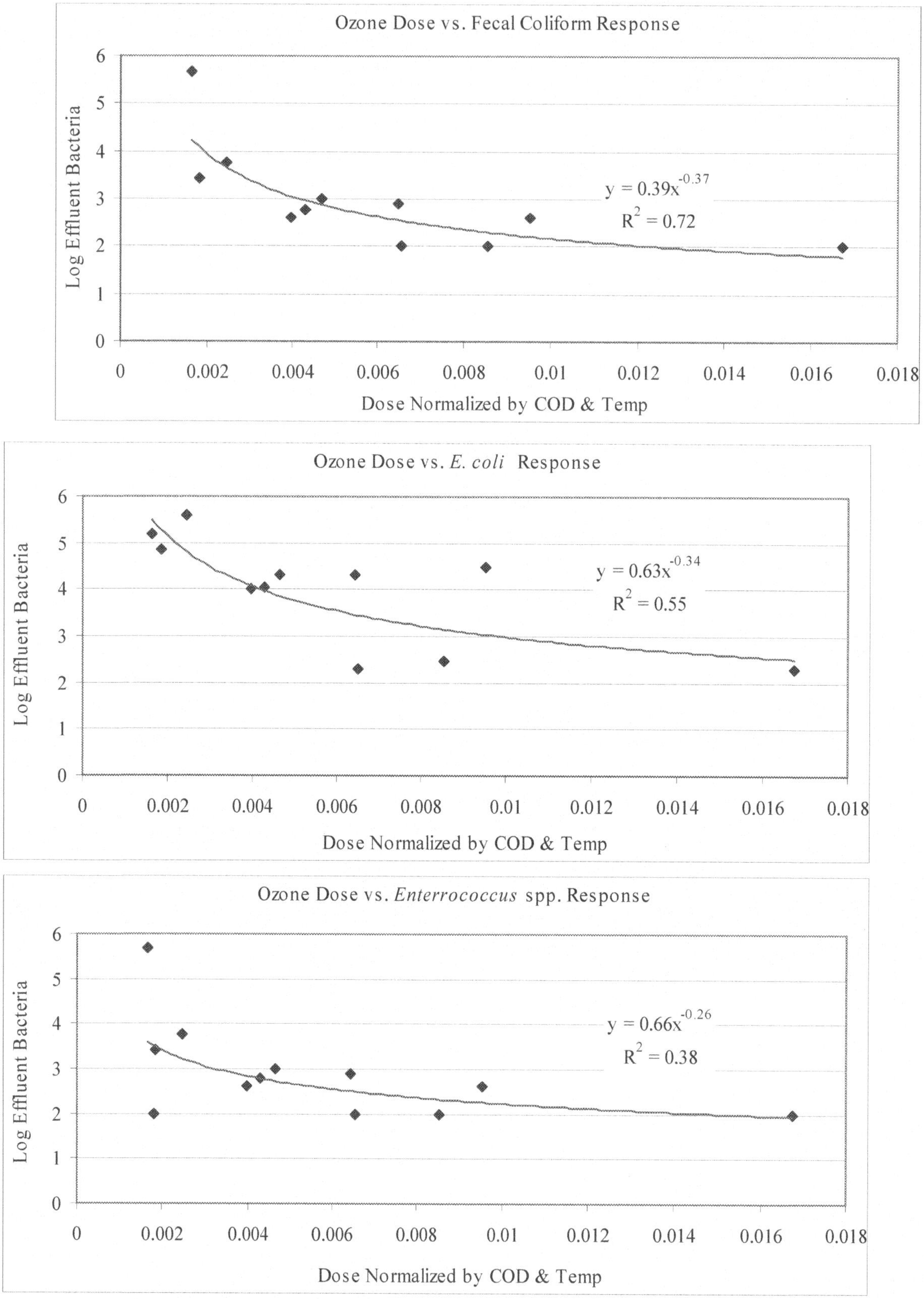

Figure 4-22. Ozone Dose-Response Curves Normalized by COD/Temperature.

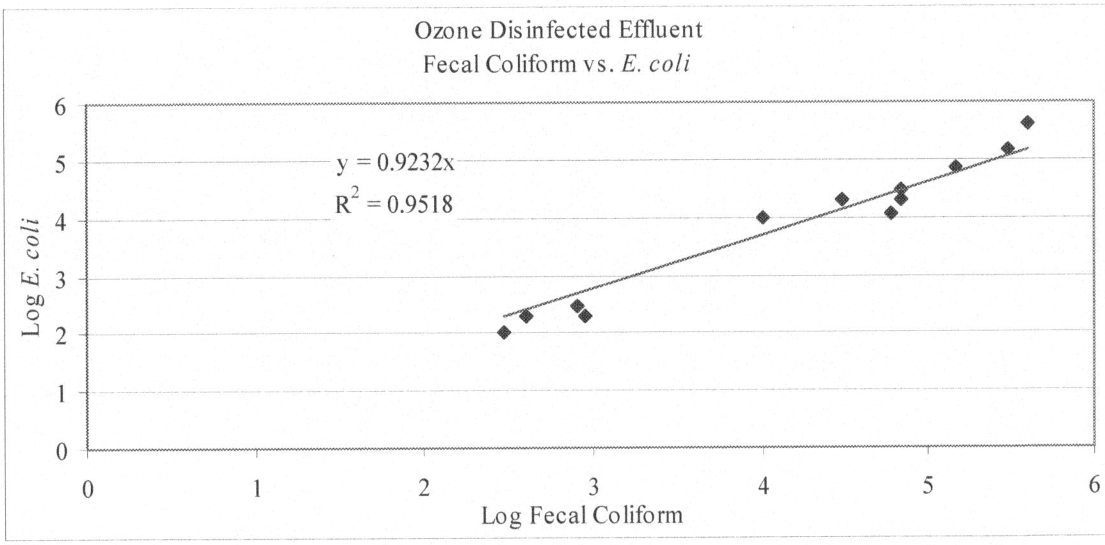

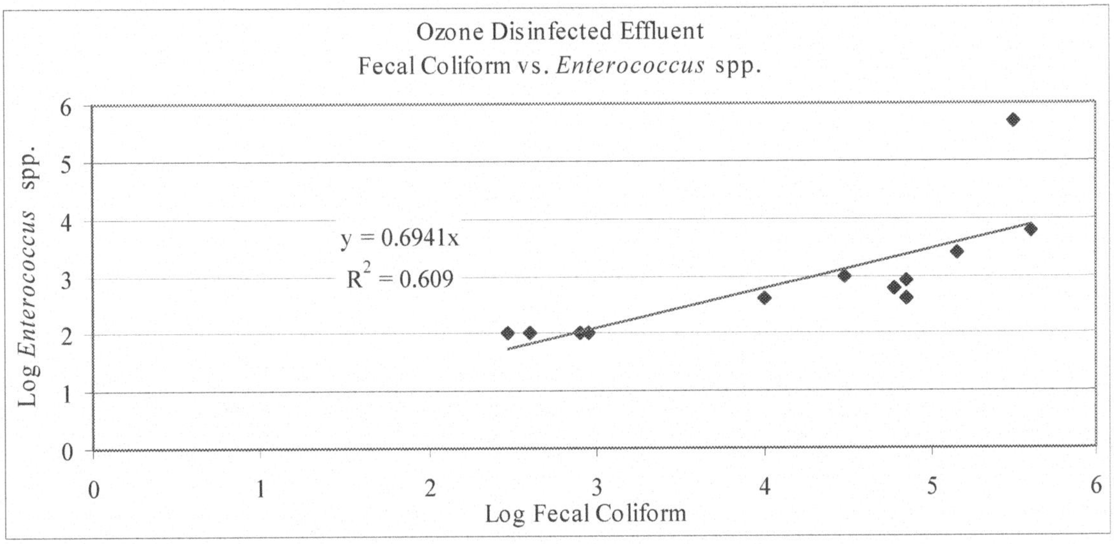

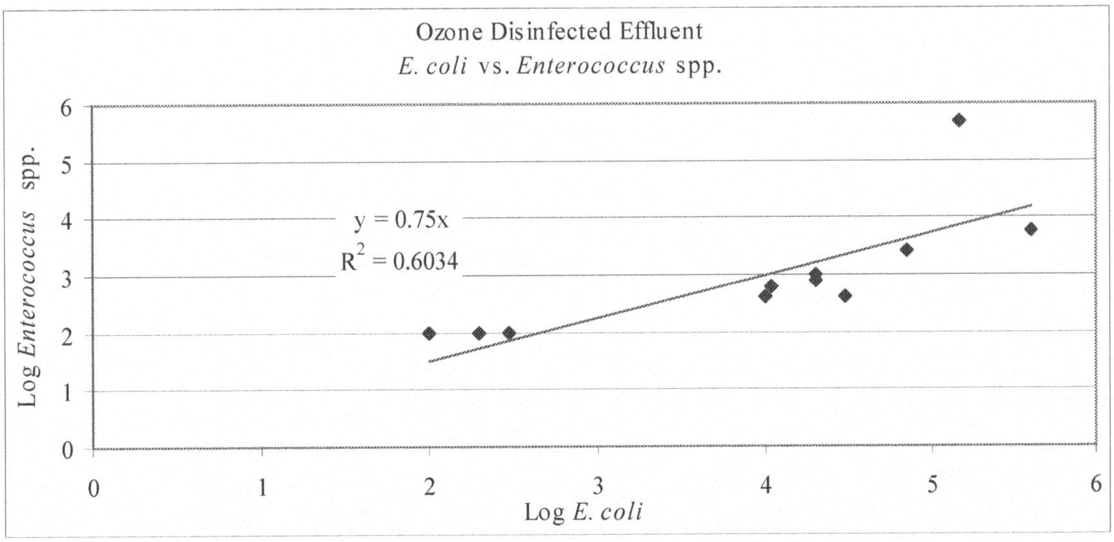

Figure 4-23. Comparison of the Bacterial in the Ozone Effluent.

4.3.5.2 Disinfection Residuals and By-products

Effluent from the ozonation disinfection system was sampled and tested for ozone residual, formaldehyde, and HAAs. Ozone residual was not detected in six of the eight measured events. The formaldehyde and HAA results are similar to measured influent values and show no significant increase. The differences between the influent and effluent disinfection residual and DBP concentrations are attributed to analytical variability. The results are provided in Table 4-16. In light of these findings, plots are not provided for ozone residual, formaldehyde, and HAAs as they were for chlorination and chlorine dioxide.

Table 4-16. Ozonation Effluent Disinfection Residual and DBPs Data.

Event	1	2	3	4	5	6	7	8	9
Dose (mg/L)	No Data	13	13	26	5	11	35	21	18
Residual Ozone (mg/l)	↓	0.13	<0.03	<0.03	<0.03	<0.03	0.18	<0.03	<0.03
Formaldehyde (mg/l)		<1	<1	<1	<1	NM	NM	NM	NM
Dibromoacetic acid (ug/l)		<1	<1	<1	<1	<1	<1	1.4	<1
Dichloroacetic acid (ug/l)		7.8	5.3	4.1	7.6	5.1	4	7.2	5.4
Monobromoacetic acid (ug/l)		<1	<1	<1	<1	<1	<1	<1	<1
Monochloroacetic acid (ug/l)		5.2	<2	<2	<2	<2	<2	2.8	<2
Trichloroacetic acid (ug/l)		11	8.3	4	7	5.6	3.2	4.4	8.6
Acute Toxicity - Fathead Minnow		>100	>100	>100	NM	NM	NM	NM	NM
Acute Toxicity - *C. dubia*		>100	>100	>100	>100	>100	>100	>100	>100
Microtox™		>100	>100	>100	>100	>100	NM	NM	NM

Notes: No data was collected during event one due to operational problems. NM=not measured because sufficient data was collected to show lack of aquatic toxicity.

4.3.5.3 Toxicity

Measured Aquatic Toxicity

Ozone-treated effluent was not toxic to fathead minnows for each of the first four collection events in this demonstration (WET tests using the fathead minnow were only performed during the first four events). Ozone-treated effluent was also not significantly toxic to the invertebrate, *C. dubia*, for each of the nine demonstration events in this study. Consequently, the reported median lethal concentration (LC_{50}) for each species for each collection event was >100%. Microtox bacterial cultures exhibited minor reductions in luminescence on exposure to ozone-treated effluent in each of the first six collection events resulting in EC_{50}s $\geq$100%. Microtox analyses were excluded from the final three events because of the consistent and minor reductions in luminescence.

Literature Review Toxicity Findings

No DBPs were measured nor was the ozone-treated effluent toxic to fathead minnows, *C. dubia*, or Microtox bacterial cultures. Therefore no literature review for specific residuals or DBPs related to ozonation was conducted.

4.3.6 Ultraviolet Light

Disinfection with UV was tested on 11 wet weather flow samples (event six was run under three different disinfection doses). The dosing concentration ranged from 65 to 220 mW*sec/cm^2. Flows were held constant during a test but varied from one test to another (1 to 7.5 GPM). The range of percent reductions and average log reductions for the three types of bacteria tested in the demonstration are summarized in Table 4-17. The low percent reduction for fecal coliform and *E. coli* were skewed downward by data from event nine; all other events had percent reductions above 99%. UV was equally effective at inactivating fecal coliform and *Enteroccoccus* spp. and least effective at inactivating *E. coli*.

Table 4-17. Bacterial Data Summary for UV Disinfection.

Bacterial Type	Initial Count Range (CFU/100 mL)	% Reduction Range	Avg. Log Reduction
Fecal coliform	1.4E5 to 1.6E7	90.833 to 99.996	3.0
E. coli	3.0E4 to 1.2E7	83.333 to 99.998	2.1
Enterococcus spp.	2.9E3 to 4.8E6	96.552 to 99.982	2.9

4.3.7 Dose-Response Relationships

Scatter plots were developed for UV dose versus log of effluent bacterial concentrations and log bacterial reductions for each of the three bacterial groups. Linear, logarithmic, and power regressions were tested against each data set, and the strongest relationship was used for calculating the coefficient of determination shown on each plot. The strongest relationship was usually observed using a power regression equation.

Similar to the chemical disinfection processes, UV dose normalized by COD and temperature offered an effective means of analyzing the performance of the UV system. All three bacterial indicators showed similar correlation with dose normalized by COD and temperature. As with the chemical disinfectants, the more specific bacteria (*Enterococcus* spp.) had the stronger correlations.

Similar to the chemical disinfection results, a two-phase disinfection effect is seen and is assumed to be due to the presence of harbored bacteria (WERF, 1996). This may help explain some of the scatter in the data. Inaccuracies in the dose measurement may be another factor contributing to the lower dose-response correlation factors. Calculating UV dose is dependent upon a variety of dynamic variables, and therefore accurate dose data are difficult to obtain. Dose is calculated by adjusting for a reduction of the intensity based upon the measured light transmittance. Similar to other disinfectants, UV effectiveness is dependent upon influent water quality characteristics, namely fluid transmittance and turbidity.

Based on a typical wet weather flow at METRO, the normalized dose-response curves indicate that it will require approximately 24% less UV dose to achieve a 1,000 CFU/100 mL for *Enterococcus* spp. as compared to achieving 1,000 CFU/100 mL for fecal coliform. This is likely the result of the higher influent concentrations of fecal coliform. To meet the maximum value criteria for each indicator organism (4,274 CFU/100 mL for fecal coliform, 2,507 CFU/100 mL for *E. coli*, and 656 CFU/100 mL for *Enterococcus* spp.), it requires approximately 24% more UV dosage to inactivate *Enterococcus* spp. as compared to fecal coliform. The UV dose necessary to achieve bacterial concentrations of 1,000 CFU/100 mL and the maximum value criteria are summarized in Table 4-18. Again, the values are calculated from the dose-response relationship normalized for a COD of 200mg/L and temperature of 20°C.

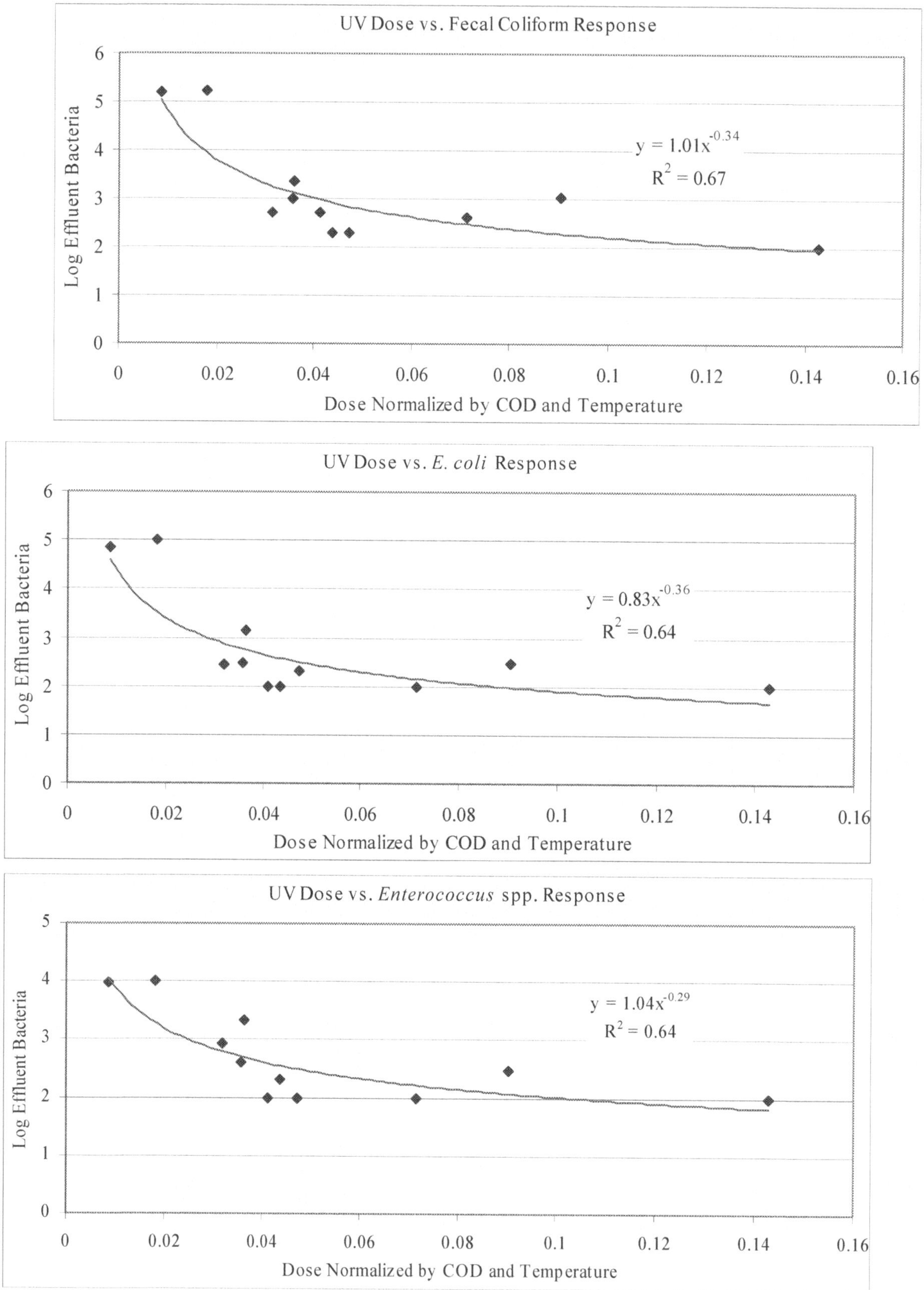

Figure 4-24. UV Dose-Response Curves Normalized by COD/Temperature.

Table 4-18. UV Dose Required to Meet Bacterial Criteria.

Parameter	UV Dose to Achieve Maximum Value Criteria	Maximum Value Criteria	UV Dose to Achieve 1,000 CFU/100 mL
Fecal coliform	110 mWsec/cm^2	4,274 CFU/100 mL	161 mWsec/cm^2
E. coli	103 mWsec/cm^2	2,507 CFU/100 mL	129 mWsec/cm^2
Enterococcus spp.	136 mWsec/cm^2	656 CFU/100 mL	119 mWsec/cm^2

Note: The above table uses the correlations based upon dose normalized by COD and temperature.

Relationships between fecal coliform, *E. coli*, and *Enterococcus* spp. in the disinfected effluent are shown in Figure 4-25. The UV-disinfected effluent shows strong correlations between all three bacterial indicators, which indicates that fecal coliform data from a UV disinfected effluent can be transformed into *E. coli* and *Enterococcus* spp. with some level of certainty. Additionally, comparing the slopes of the lines from the influent charts (Figure 4-2) to the effluent charts (Figure 4-25) provides an indication of the effectiveness of UV to inactivate the different bacterial concentrations. Based on changes to the slope of the lines, UV was most effective at inactivating *Enterococcus* spp., which was likely because fecal coliform and *E. coli* had the highest challenge concentrations.

4.3.7.1 Disinfection Residuals and By-products

Effluent from the UV disinfection system was sampled and tested for ozone, formaldehyde, and HAAs. The results are very similar to measured influent values and show no significant increase. Differences between the influent and effluent numbers are attributed to analytical variability. The results are provided in Table 4-19.

Table 4-19. UV Effluent Disinfection Residual and DBP Data.

Event	1	2	3	4	5	6	7	8	9
Dose (mW*s/cm2)	200	67	130	92	90	73	190	98	65
Ozone (mg/l)	<0.03	<0.03	<0.03	<0.03	<0.03	<0.03	<0.03	<0.03	<0.03
Formaldehyde (mg/l)	No Data	<1	<1	<1	No Data	NM	NM	NM	NM
Dibromoacetic acid (ug/l)	<1	<1	<1	<1		<1	<1	<1	<1
Dichloroacetic acid (ug/l)	1.7	8.1	6	3.9		4.7	3.6	7.7	6.7
Monobromoacetic acid (ug/l)	<1	<1	<1	<1		<1	<1	<1	<1
Monochloroacetic acid (ug/l)	<2	<2	<2	<2		<2	<2	<2	<2
Trichloroacetic acid (ug/l)	2.7	4.6	9.6	3.9		6.4	2.7	6.9	6.1
Acute Toxicity - Fathead Minnow	>100	>100	>100	>100		NM	NM	NM	NM
Acute Toxicity - *C. dubia*	>100	>100	>100	>100		>100	>100	>100	>100
Microtox™	>100	>100	>100	>100		>100	NM	NM	NM

Notes: No data was collected during event five due to sampling error. NM=not measured because sufficient data was collected to show lack of aquatic toxicity.

4.3.7.2 Toxicity

<u>Measured Aquatic Toxicity</u>

UV-treated effluent was not toxic to fathead minnows in demonstration the events (two through four) WET tests were performed that UV-treated effluent was also not significantly toxic to the invertebrate, *C. dubia*, for each of the nine collection events in this demonstration and resulted in LC_{50}s for each species of >100%. Microtox bacterial cultures exhibited consistent and minor reductions in luminescence on exposure to UV-treated effluent in each of the first six collection events, resulting in EC_{50}s >100% and were subsequently excluded from the final three events.

Literature Review Toxicity Findings

No DBPs were measured nor was the UV-treated effluent toxic to fathead minnows, *C. dubia*, or Microtox bacterial cultures. Therefore, no literature review for specific DBPs related to UV was conducted.

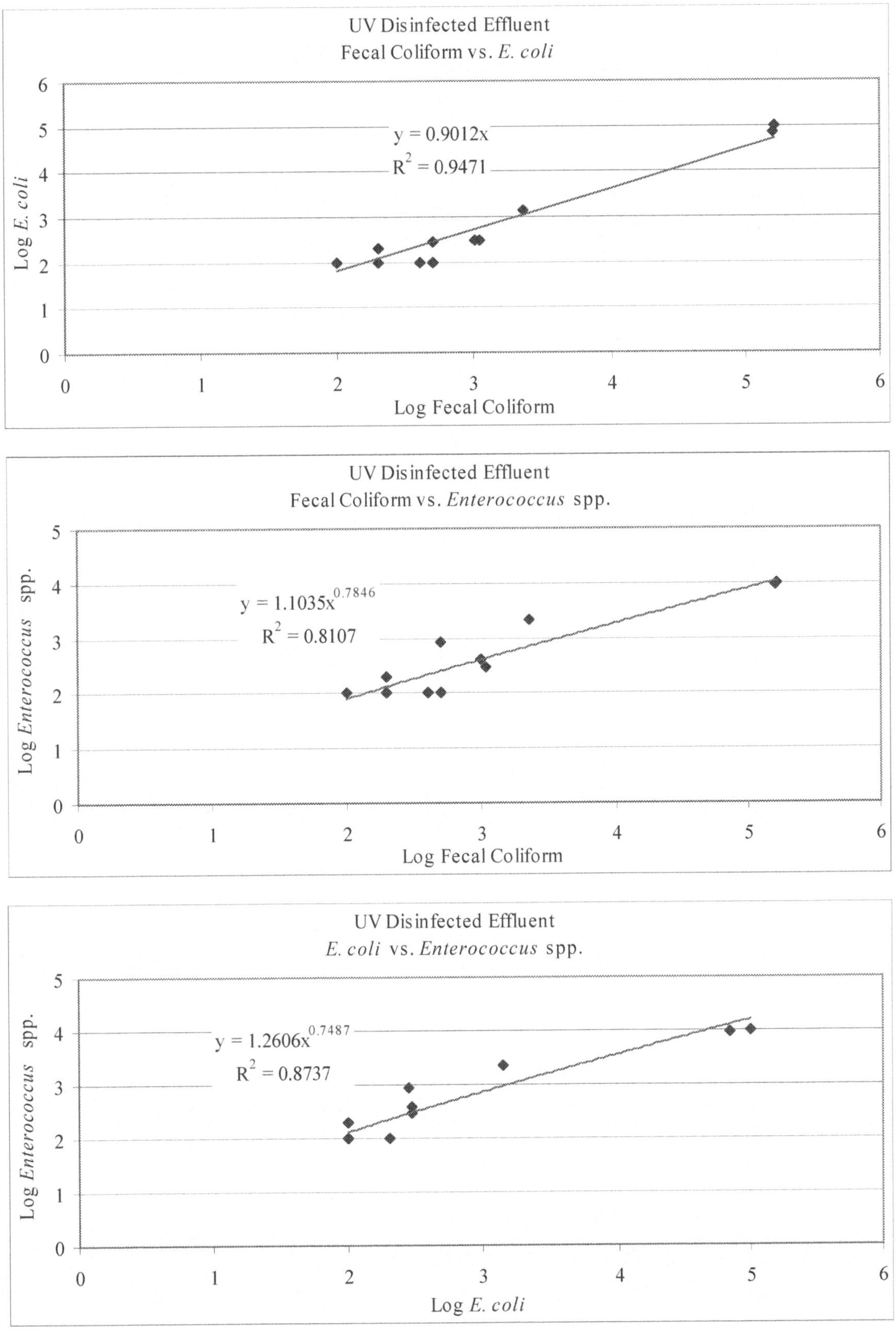

Figure 4-25. Comparison of the Bacterial in the UV Effluent.

CHAPTER 5.0

APPLICATIONS FOR FULL-SCALE FACILITIES

5.1 Site-Specific Application of Findings

Design of wet weather flow disinfection facilities should be based on the specific characteristics and dynamics of the particular wet weather flow discharge and receiving water in question. Although a WWTP influent under wet weather conditions is somewhat representative of the wet weather flow discharges within the collection system, it does not take the place of specific water quality data from a wet weather flow discharge. However, the WWTP influent does provide the benefit of the sustained flows from one or more wet weather flow discharges as the storm moves across the collection system. Therefore, for this disinfection demonstration WWTP influent was used. Nevertheless, the results from a comprehensive wet weather flow monitoring program should be used to adjust data collected at the WWTP. For example, for this demonstration project, the combined sewage that occurs at the Syracuse METRO WWTP during wet weather has fairly high COD with relatively little variability as compared to combined sewage in the upper collection system, which has dilute COD levels and highly variable pollutant loadings.

In addition to characterizing the wet weather flow, the relationship of the discharge to the receiving water needs to be defined in order to more accurately assess the impact that the bacteria and disinfection by-products will have on the environment. For instance, the hydrodynamics and aquatic biology of the receiving water may require an environmentally sensitive disinfection technology like UV, whereas a large receiving waterbody may be less sensitive and allow chemical disinfection. An alternative may include chemical disinfection with residual removal.

Before considering any site specific application of the results from this disinfection demonstration it is important to note that the data and results presented in Chapter 4.0 are based on the following conditions:

- The subject wet weather flow was collected at a WWTP, which will generally have higher concentrations of pollutants, with the possible exception of grit, than CSOs.
- Disinfection residual and DBP concentrations do not account for receiving water dilution; and
- Disinfection doses were based on meeting the draft EPA guidance values (U.S. EPA, 2002) in the effluent (i.e., no dilution):

  -fecal coliform of 4,274 CFU/100 mL
  -*E. coli* of 2,507 CFU/100 mL
  -*Enterococcus* spp. of 656 CFU/100 mL

5.2 Application of Findings to Onondaga County Case Study

Adjustments were made to the disinfection demonstration findings based on specific conditions in Onondaga County, namely, adjustment of disinfectant dose based on actual CSO water quality and adjustment of DBP concentrations based on receiving water dilutions. Additionally, a bacterial criterion for fecal coliform of 200 CFU/100 mL that was previously established under a consent order was used instead of the draft U.S. EPA guidance values (4,274 CFU/100mL).

5.2.1 Bacterial Criterion

During the years 1999-2000, Onondaga County modeled bacteria impacts resulting from wet weather flow discharges to the receiving waters. This modeling effort used as a criterion a fecal coliform concentration not to exceed 200 CFU/100 mL in Onondaga Lake. This criterion was approved by the regulatory agency and served as a basis for the consent order. The receiving waters are described in more detail in Section 5.2.3.

Based on receiving water characteristics, which included bacterial reductions due to natural die-off and sedimentation, a fecal coliform concentration of 200 CFU/100 mL in Onondaga Lake translated into a 2.6-log reduction of fecal coliform at the major CSO discharges. A 2.6-log reduction of fecal coliform further translated into an end-of-pipe concentration in the order of 2,000 CFU/100 mL. Coincidentally, the data and results presented in Chapter 4.0 were based on an end-of-pipe fecal coliform concentration in the same order of magnitude of 4,274 CFU/100 mL.

5.2.2 Water Quality Adjustments

The wet weather flow at the Syracuse METRO WWTP has fairly high concentrations of constituents as compared to the wet weather flow discharges from combined sewage in the upper collection system. Based on historical data, the average COD concentration measured at the Newell Street overflow averaged 95 mg/L as compared to the METRO average of 182 mg/L measured during this demonstration. Temperatures were similar at both locations. Therefore, by using the dose-response curve normalized by COD and temperature, a targeted dose to achieve a desired fecal coliform concentration can be selected. An example of this process is presented below.

Figure 5-1 shows the fecal coliform versus dose normalized by COD and temperature for the chlorination disinfection process as presented in Chapter 4.0. This dose-response curve can be used to determine the targeted disinfection dose. First, the targeted effluent fecal coliform concentration must be defined; in the case of Onondaga County the fecal coliform target is 2,000 CFU/100 mL, or 3.3 log. The log of the fecal coliform concentration is the y value, which translates into an x value of 0.0046 when using the dose-response curve. The x value of 0.0046 is the targeted chlorine dose normalized by COD and temperature. The normalized dose is then corrected by the site-specific water quality concentrations. In the case of the Newell Street CSO, the targeted dose can be calculated by multiplying 0.0046 by 95 and then by 20, which yields an average chlorine dose of approximately 9 mg/L. Following these same procedures, site specific average doses for chlorine dioxide, ozonation, and UV were calculated and the results were as follows:

- Chlorine dioxide = 4 mg/L
- Ozonation = 6 mg/L
- UV = 60 mWsec/cm^2

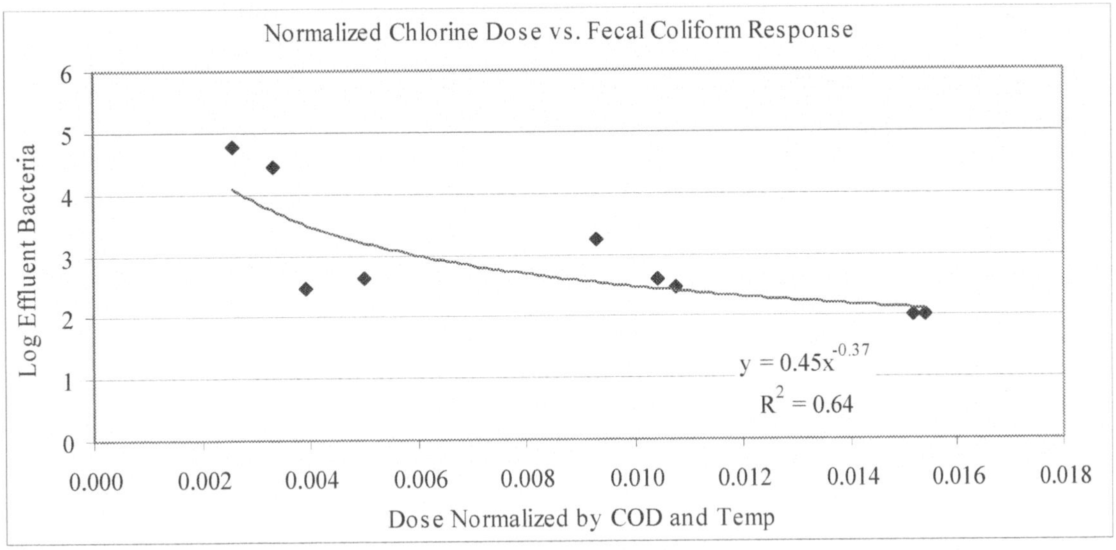

Figure 5-1. Chlorine Dose-Response Curves Normalized by COD/Temperature.

5.2.3 Receiving Water Dilution Adjustments

The collection system tributary to the METRO WWTP is a combined system. The CSOs are located primarily along Onondaga Creek which flows into the southern end of Onondaga Lake. METRO also discharges into the southern end of Onondaga Lake.

For regulatory purposes, the extreme southern end of Onondaga Lake has been deemed a mixing zone, meaning that this section of the lake does not have as stringent water quality criteria as the lake proper. The lake proper is designated as recreational waters for swimming.

Onondaga Creek is not used for swimming; therefore, human exposure to pathogens and/or disinfection residuals and by-products in the creek would be incidental. However, aquatic exposure to residuals and by-products in Onondaga Creek represent an exposure pathway for aquatic organisms.

In light of these two different receiving waters and their different designated uses, two different approaches for estimating environmental concentrations were developed. One approach was developed for Onondaga Creek where exposure to residuals and by-products is expected for aquatic ecological receptors, and one for Onondaga Lake where exposure is expected for both human and aquatic ecological receptors. The Onondaga Creek dilution adjustment was based on the proportion of background creek flow versus CSO-discharge flow. The Onondaga Lake dilution adjustment was based on the proportion of the mixing zone volume in the extreme southern part of the lake versus the CSO-discharge volume.

5.2.3.1 Onondaga Creek Dilution Adjustments

Estimating environmental concentrations of disinfection residuals and by-products in Onondaga Creek was based on the minimum average seven-consecutive-day low flow with a 10-year recurrence interval (MA7CD10) in Onondaga Creek occurring simultaneously with the one-year recurrence wet weather event. The one-year recurrence wet weather event is the rain event in terms of both intensity and volume that occurs on average once per year in the Onondaga Lake basin. This is a conservative approach, in that the combined probability of a one-year storm occurring during a MA7CD10 stream flow is far less frequent than once in 10 years. For example, if a two-hour duration storm that is likely to occur from May through September is (Pr), and the minimum stream flow is (Ps), the joint probability (Pj)of Pr and Ps combined is calculated as:

$$Pr * Ps = Pj$$

$$\frac{1}{155\frac{hrs}{yr} * \frac{1yr}{2hrs}} * \frac{1}{8760\frac{hrs}{yr} * 10yr} = \frac{1}{6{,}789{,}000hrs} \text{ (once in 77 years).}$$

The MA7CD10 and one-year recurrence wet weather event assumptions provide a "minimum" dilution based upon which residual and by-product concentrations are calculated, which results in a conservative exposure estimate.

The following equation was developed to determine the dilution in Onondaga Creek:

$$Q_{rw}C_{rw} + Q_D C_D + Q_{ND} C_{ND} = Q_{ec}C_{ec}$$

where:

Q_{rw} =	MA7CD10 flow in the receiving water
C_{rw} =	concentration of chemical in the receiving water during MA7CD10 (i.e., background concentration)
Q_D =	wet weather flow resulting from the one-year recurrence wet weather event that is disinfected
C_D =	concentration of chemical in the disinfected wet weather flow
Q_{ND} =	wet weather flow resulting from the one-year recurrence wet weather event that is NOT disinfected
C_{ND} =	concentration of chemical in the non-disinfected wet weather flow
Q_{ec} =	estimated flow in the receiving water (MA7CD10 plus wet weather flow)
C_{ec} =	estimated environmental concentration

In the case of Onondaga Creek, the MA7CD10 is 26 cubic feet per second (CFS). The peak one-year recurrence wet weather flow is approximately 3,100 CFS; however, only a portion of this flow is disinfected. The disinfected portion of this flow is approximately 1,300 CFS, and the non-disinfected portion of this flow is approximately 1,800 CFS. It was found through the bacteria modeling (Section 5.2.1) that only a portion of the wet weather flow discharges required disinfection to meet the fecal coliform criterion in Onondaga Lake. The ratio of disinfected wet weather flow to creek flow plus disinfected and non-disinfected wet weather flow is approximately 1:2.4. Therefore, the concentrations of the residuals and by-products in the effluent of the disinfected wet weather flows can be adjusted by a factor of 0.42 to reflect the concentrations expected in Onondaga Creek. For example, an end-of-pipe concentration of 10 µg/L would translate into a concentration of 4.2 µg/L in Onondaga Creek.

5.2.3.2 Onondaga Lake Dilution Adjustments

Estimating environmental concentrations and disinfection residuals and by-products in Onondaga Lake was based on the dilution provided by the volume of the mixing zone in the extreme southern end of Onondaga Lake during the one-year recurrence wet weather event.

The following equation was developed to determine the appropriate dilution factor in Onondaga Lake:

$$VmzCmz + V_{ND}C_{ND} + V_DC_D = VecCec$$

where:

V_{mz} = volume of mixing zone

C_{mz} = concentration of chemical in the mixing zone

V_D = wet weather flow volume resulting from the one-year recurrence wet weather event that is disinfected

C_D = concentration of chemical in the disinfected wet weather flow volume

V_{ND} = wet weather flow volume resulting from the one-year recurrence wet weather event that is NOT disinfected

C_{ND} = concentration of chemical in the non-disinfected wet weather flow volume

V_{ec} = estimated volume in receiving water (mixing zone volume plus wet weather flow volume)

C_{ec} = estimated environmental concentration

In the case of Onondaga Lake, the designated mixing zone is 2,540 million gallons, and the one-year recurrence wet weather event produces a wet weather flow volume of 52 million gallons. Of the 52 million gallons, 21 million gallons is disinfected, and the remaining 31 million gallons is not disinfected. The ratio of the disinfected wet weather flow volume to the total volume in the mixing zone volume (including the disinfected and non-disinfected wet weather flow volumes) is approximately 1:123. Therefore, when considering concentrations of disinfection residuals and by-products in Onondaga Lake, a dilution factor of 0.008 should be applied to the end-of-pipe concentrations measured during this disinfection demonstration. For example, an end-of-pipe concentration of 10 µg/L would translate into a concentration of 0.08 µg/L in Onondaga Lake.

5.2.4 Estimated Environmental Concentrations

Effluent concentrations of disinfection residuals and DBPs measured during the disinfection demonstration were extrapolated to predict estimated environmental concentrations (EECs) using the adjustment factors presented in Sections 5.2.2 and 5.2.3. This extrapolation process used end-of-pipe concentrations of residuals and DBPs resulting from the average doses presented in Section 5.2.2. This approach is especially important when considering residual and DBP concentrations with respect to human and ecological risk assessments as presented in Section 5.4. Tables 5-1 through 5-3 present the end-of-pipe concentrations, the estimated Onondaga Creek concentrations, and the estimated Onondaga Lake concentrations, respectively. Only chemicals that increased in concentration as a result of the disinfection processes are presented in Tables 5-1 and 5-2.

Table 5-1. Targeted Dose and Resulting Residuals and DBPs in the Disinfected Effluent (End of Pipe).

Parameter/Technology	Chlorine	Chlorine Dioxide	Ozonation	UV
Dose[1] (mg/L)	9	4	6	60[2]
Residual w/o Reducing Agent (mg/L)	2.1	0.04	ND[3]	N/A[4]
Chlorate (mg/L)	<1	2.2	NM[5]	NM
Chlorite (mg/L)	ND	14	NM	NM
Chloroform (µg/L)	2.3	ND	NM	NM
Monochloroacetic acid (µg/L)	ND	4.1	NM	NM
Dichloroacetic acid (µg/L)	10	28	NM	NM
Trichloroacetic acid (µg/L)	8.2	8.3	NM	NM

[1] Dose required to achieve 2,000 CFU/100 mL fecal coliform in effluent

[2] Dose units mWsec/cm^2

[3] Measured, but not detected [4] Not applicable [5] Not measured

Table 5-2. Targeted Dose and Resulting Residuals and DBPs in Onondaga Creek.

Parameter/Technology	Chlorine	Chlorine Dioxide	Ozonation	UV
Dose[1] (mg/L)	9	4	6	60[2]
Residual w/o Reducing Agent (mg/L)	0.88	0.02	ND[3]	N/A[4]
Chlorate (mg/L)	<0.5	0.92	NM[5]	NM
Chlorite (mg/L)	ND	5.9	NM	NM
Chloroform (μg/L)	1.0	ND	NM	NM
Monochloroacetic acid (μg/L)	ND	1.7	NM	NM
Dichloroacetic acid (μg/L)	4.2	12	NM	NM
Trichloroacetic acid (μg/L)	3.4	3.5	NM	NM

[1] Dose required to achieve 2,000 CFU/100 mL fecal coliform in effluent
[2] Dose units mWsec/cm^2
[3] Measured, but not detected [4] Not applicable [5] Not measured

Table 5-3. Targeted Dose and Resulting Residuals and DBPs in Onondaga Lake.

Parameter/Technology	Chlorine	Chlorine Dioxide	Ozonation	UV
Dose[1] (mg/L)	9	4	6	60[2]
Residual w/o Reducing Agent (mg/L)	0.017	0.0003	ND[3]	N/A[4]
Chlorate (mg/L)	<0.01	0.018	NM[5]	NM
Chlorite (mg/L)	ND	0.11	NM	NM
Chloroform (μg/L)	0.018	ND	NM	NM
Monochloroacetic acid (μg/L)	ND	0.032	NM	NM
Dichloroacetic acid (μg/L)	0.081	0.22	NM	NM
Trichloroacetic acid (μg/L)	0.066	0.066	NM	NM

[1] Dose required to achieve 2,000 CFU/100 mL fecal coliform in effluent
[2] Dose units mWsec/cm^2
[3] Measured, but not detected [4] Not applicable [5] Not measured

5.3 Screening-level Risk Assessment

5.3.1 General Approach

The generation of disinfection residual and DBPs has become a concern for various disinfection technologies. As presented in Table 4-2, Section 4.3.2, several disinfection residuals and DBPs were measured during the disinfection demonstration. If there was a significant increase in the concentration of either a residual or DBP in the effluent of a disinfection pilot, then a screening-level risk assessment was performed on the specific residual or DBP in order to estimate whether the effluent discharge may cause adverse effects in human or aquatic organisms.

In addition to this screening-level risk assessment, aquatic toxicity was measured as part of the disinfection demonstration. Aquatic toxicity was measured in the effluent of each of the disinfection pilots, and the data are presented as part of the disinfection demonstration results under each disinfection technology. These data are summarized in this section as they relate to the screening-level risk assessment.

The screening-level risk assessment included the following steps, which are consistent with the general risk-assessment paradigm established by U.S. EPA (U.S. EPA, 1989):

- Hazard Identification
- Exposure Assessment
- Toxicity Assessment
- Risk Characterization

Through the hazard identification literature review (Chapter 8.0), likely residuals and DBPs were identified. Once identified, these residuals and DBPs were further researched through the literature to obtain toxicity information relevant for assessing risks to human health and the aquatic ecosystem. These findings were used to establish risk-based concentrations (RBCs) as toxicity reference values. No new toxicity reference values were derived during this screening-level risk assessment process; only toxicity values found in the literature were used. Assessing exposure included determining what concentrations of these residuals and DBPs were expected to occur in the environment as well as determining likely exposure pathways. For all the chemicals with sufficient toxicity and exposure data, EECs were compared to the RBC values. Chemicals with EECs that exceeded the RBC were characterized as chemicals of potential concern (COPC) because concentrations that exceed risk-based thresholds may pose unacceptable risks to humans or the aquatic ecosystem. Since this was a screening-level assessment, COPC classification means that additional study may be warranted.

5.3.2 Hazard Identification

In general, a chemical may be considered for a screening-level risk assessment if there is evidence that the chemical may be present in the environment, that exposures may occur by some route, and that an adverse health effect may result from the exposure. For this project, hazard identification begins with a description of the chemicals that may be generated by each of the disinfection processes. A subset of these chemicals has been identified to conduct a screening-level risk assessment based on evidence of potential adverse health effects to humans or aquatic ecological receptors. A literature review was conducted to evaluate the epidemiological and toxicity study information available to assess potential risks to humans and aquatic organisms exposed to residuals and DBPs (Chapter 8.0). A range of adverse health effects may be considered for both humans and aquatic organisms.

A broad distinction in health effects is made in human health risk assessment. Chemicals may be evaluated for potential cancer and/or noncancer health effects, and regulatory decisions may be determined for either endpoint. For example, different RBCs may be established for the same chemical, depending on concerns for exposures that may result in cancer or noncancer health effects. Both the epidemiological literature and the mammalian toxicology literature were considered in the review of information relevant to human health. For this report, no attempt was made to derive human health toxicity values or protective concentration levels. Regulatory standards are set based on available data and extensive scientific review. For example, a primary source of human health toxicity values is the Integrated Risk Information System (IRIS) web site (http://www.epa.gov/iris/), which also provides background information and explanations of the methods used to derive the values. Health assessment information on a chemical is posted on IRIS only after undergoing a comprehensive review of chronic toxicity data by U.S. EPA health scientists from several program offices, the Office of Research and Development, and outside experts. Note that the IRIS review process often excludes many studies on the basis of quality or representativeness. *In this screening-level assessment, only those chemicals for which regulatory values have been established were used.*

For aquatic organisms, the health effects of concern depend on the definition of the ecological risk assessment endpoint, which is the ecological level at which the chemical may have an adverse impact (e.g., individual organism, population, community, or ecosystem), and the nature of the effect (e.g., mortality, reduced fecundity, impaired growth) (U.S. EPA, 1992a). For this demonstration project report, the focus for aquatic ecotoxicity is on identifying chemical concentrations that have demonstrated a moderate likelihood of mortality (or immobility, an indicator of mortality for many invertebrates) to a fraction of a population of at least one species of aquatic organisms that is generally considered to be an indicator species for the health and function of freshwater or marine ecosystems. For this project, toxicity reference values were obtained from the literature in the form of raw toxicological data. The lowest available LC_{50} for mortality was chosen as the toxicity reference value in order to be protective in a screening-level assessment.

5.3.3 Exposure Assessment

For this disinfection demonstration project, exposure pathways were defined in terms of human exposure and aquatic organism exposure, and were based on site-specific characteristics of the receiving water in the study area (Onondaga County, NY).

As presented in Section 5.2.3, the collection system tributary to the METRO wastewater treatment plant is primarily a combined system. The CSOs are located primarily along Onondaga Creek, which flows into the southern end of Onondaga Lake. The lake proper is designated as recreational waters for swimming; therefore, human exposure to disinfection residuals and DBPs need to be considered. Onondaga Creek and the extreme southern end of the lake are not used for swimming; therefore, human exposure to disinfection residuals and DBPs in the creek would be incidental. However, aquatic exposure to residuals and DBPs in Onondaga Creek represent an exposure pathway for fish and invertebrates.

Human exposure to wet weather flow discharges is expected to occur in Onondaga Lake due to full body contact such as swimming. Exposure scenarios associated with swimming include ingestion of water, inhalation, and dermal contact. Equations for calculating the chronic daily intake (CDI) for each pathway is given below.

Incidental Ingestion of Water

$$CDI = \frac{C_{water} \times CF_1 \times IR_{water}}{BW} \times \frac{EF}{365} \times \frac{ED}{AT}$$

Inhalation

$$CDI = \frac{C_{air} \times IR_{air}}{BW} \times \frac{t_{event}}{24} \times \frac{EF}{365} \times \frac{ED}{AT}$$

Dermal Contact

$$CDI = \frac{DA_{event} \times SA \times EV}{BW} \times \frac{EF}{365} \times \frac{ED}{AT}$$

where:

CDI	=	chronic daily intake (mg/kg-day)
C_{water}	=	chemical concentration in water (µg/L)
C_{air}	=	chemical concentration in air (mg/m^3)
IR_{water}	=	ingestion rate of water (L/day)
IR_{air}	=	inhalation rate (m^3/day)
EF	=	exposure frequency (days/year, divided by 365 days/year)
ED	=	exposure duration (years)
t_{event}	=	exposure time (hr/day, divided by 24 hours/day for inhalation)
BW	=	body weight (kg)
AT	=	averaging time (days)
CF_1	=	conversion factor 1 (0.001 mg/µg)
SA	=	skin surface area (cm^2)
EV	=	number of events per day
DA_{event}	=	absorbed dose per event (mg/cm^2-event)

and for organic chemicals DA_{event} is calculated as follows:

if $t_{event} \le t^*$, *then*

$$DA_{event} = 2 \times FA \times K_p \times C_{water} \times \sqrt{\frac{6\tau_{event} \times t_{event}}{\pi}}$$

if $t_{event} > t^*$, *then*

$$DA_{event} = FA \times K_p \times C_{water} \left[\frac{t_{event}}{1+B} + 2\tau_{event} \left(\frac{1+3B+3B^2}{(1+B)^2} \right) \right]$$

and for inorganic chemicals DA_{event} is calculated as follows:

$$DA_{event} = K_p \times C_{water} \times t_{event}$$

where:

FA	=	fraction absorbed (unitless)
K_p	=	permeability constant (cm/hr)
τ_{event}	=	lag time per event (hr/event)
t*	=	time to reach steady state (hr) = 2.4 τ_{event}
π	=	pi (unitless)
B	=	dimensionless ratio of the permeability coefficient of a compound through the stratum corneum relative to its permeability coefficient across the viable epidermis

5.3.3.1 Exposure Scenarios

Human exposure to wet weather flow discharges is expected to occur in Onondaga Lake due to full body contact such as swimming. Exposure scenarios associated with swimming include ingestion of water, inhalation, and dermal contact.

The relative contributions of each exposure pathway to the total chronic daily intake of a chemical depends on several factors, including the concentration and properties of the chemical released into the environment, as well as the activity patterns, contact rates, and other characteristics of the exposed population. In a screening-level risk assessment, the objective is to focus the risk calculations on the factors that are likely to contribute the most to the risk assessment.

The conceptual model for the screening-level assessment includes the assumption that the treated water containing residuals and DPBs may be used for recreation but not as a main source of residential drinking water. Nevertheless, it is possible that individuals will consume small amounts of water during swimming. Toxicity data available to assess the potential adverse health effects associated with exposures are most commonly available for oral exposure pathways. Therefore, if concentration data are available to assess exposure, then the chronic daily intake can be calculated using the above equation for incidental ingestion, and the result can be combined with available toxicity information in order to estimate cancer and noncancer health risks.

The risk of exposure in inhalation is unlikely to be a dominant pathway for chemicals with a low potential for volatilization. Volatilization from a waterbody is controlled by the properties of the chemical, the properties of the waterbody and the properties of the air above the waterbody. Chemicals that have a Henry's law constant of less than $3x10^{-7}$ atm-m^3/mol are generally considered nonvolatile from waterbodies since they tend to volatilize more slowly than water itself. Compounds that exist as an anion or cation are also considered nonvolatile since ionic species have infinitely low vapor pressures and Henry's law constants. For this reason, chlorate and chlorite ions are considered nonvolatile, and since the pK_a of the chlorinated species of acetic acid are low ($pK_a < 4$), these compounds will exist primarily as anions in water and will not volatilize.

Of the residuals and DPBs that increased in concentration due to the disinfection processes, chlorine, chlorine dioxide, and chloroform were considered to be potentially volatile. The volatilization half-life of chlorine, chlorine dioxide, and chloroform were estimated using the two-layer approach of Liss and Slater (1974) and discussed in Lyman et al. (1990). Using appropriate Henry's law constants, a stream depth of two meters, and assuming a water velocity and wind speed of approximately 1 m/s, the volatilization half-life of the chemicals (listed in Table 5-4) range from about 14 to 19 hours.

Table 5-4. Volatilization Half-life in a Model Stream.

Chemical	Henry's Law constant ($atm\text{-}m^3/mol$)	Half-Life (hours)
Chlorine	0.0117	14
Chloroform	0.00367	19
Chlorine dioxide	0.024	14

To calculate the concentration in air needed to estimate inhalation exposure, the swimming area of the stream was estimated as 1,000 m^2 (100 m long, 10 m wide) and the volume of water in the swimming area was 2000 m^3 (100 m long, 10 m wide, 2 m depth). For convenience, the breathing zone above the waterbody was estimated as having the same dimensions. It was further assumed that the chemical was well mixed throughout the volume of the model pond. The initial concentrations in water are based on the estimated concentrations in Onondaga Lake (see Table 5-3) for chlorine (17 μg/L), chlorine dioxide (0.3 μg/L), and chloroform (0.018 μg/L).

Ignoring the flow of water, but considering a constant wind velocity of 1 m/s, the advection half-life of any chemical in the air above the swimming area is approximately 12 minutes. Adjusting the air levels for this loss mechanism removes a portion of the chemical from the relatively small breathing zone. In each case, the peak exposure occurs at about 15 minutes and decreases rapidly, approaching 0 by 5 hours. Therefore, these data can be viewed as a conservative estimate since no degradation losses of the chemical have been assumed in either the air or water compartments. For a screening-level risk assessment, the exposure concentration in air can be determined from these peak concentrations: chlorine (0.085 mg/m^3), chlorine dioxide (0.0015 mg/m^3), and chloroform ($6.7\text{x}10^{-5}$ mg/m^3).

U.S. EPA guidance (RAGS Vol. 1, Part E; U.S. EPA, 2004a) recommends including the dermal exposure route in the risk calculation when the dermal route would pose more than 10% of the ingested dose. For most chemicals, including all of the DBPs identified as COPCs in this report, chronic daily intake via dermal contact is expected to be low, although if ingestion is restricted to incidental ingestion of water during swimming (i.e., residential drinking water is not included), then it is difficult to determine if the dermal route will be less than 10% of the ingested dose.

5.3.3.2 Parameter Values for Calculating Chronic Daily Intake (CDI)

CDIs were calculated for oral, inhalation, and dermal exposure pathways using the equations presented above and the input values given in Table 5-5. These inputs are based on

U.S. EPA guidance for exposure assessment (U.S. EPA, 1997a; U.S. EPA, 2001) as well as professional judgment. Age-specific parameter estimates were considered in order to determine if cancer or noncancer risks might be significantly higher for children than adults. For the EECs presented in this screening-level assessment, age-specific estimates of CDI would not change the conclusions, so values specific to adults are presented for simplicity.

Table 5-5. Input Parameter Values Used to Calculate CDI.

Exposure Variable	Units	Value	Reference
Ingestion rate, water	L/hr	0.05	U.S. EPA, 1989
Exposure time	hr/day	3.0	Judgment
Body weight	kg	70	U.S. EPA, 1997a
Exposure frequency	days/year	7	U.S. EPA, 1989
Exposure duration	year	30	U.S. EPA, 1989
Averaging time, cancer	year	70	U.S. EPA, 1989
Averaging time, noncancer	year	Equal to ED	U.S. EPA, 1989
Inhalation rate	m^3/hour	2.0	U.S. EPA, 1997a
Surface area	cm^2	18000	U.S. EPA, 2001
Fraction absorbed	unitless	1.0	U.S. EPA, 2001
Permeability coefficient	cm/hr	chemical-specific	U.S. EPA, 2001
Lag time per event	hr/event	chemical-specific	U.S. EPA, 2001
B (Lipophilic Property)	unitless	chemical-specific	U.S. EPA, 2001

5.3.3.3 Results of Chronic Daily Intake Calculations for Human Health

CDIs for each chemical and each exposure pathway are summarized in Table 5-6 for noncancer risks and Table 5-7 for cancer risks. The CDIs presented in Tables 5-6 and 5-7 are based on the estimated environmental concentrations in Onondaga Lake (Table 5-3) resulting from both the chlorination and chlorine dioxide disinfection processes. In some cases, there are different CDIs presented for the same chemical because different environmental concentrations were estimated for the chlorination and chlorine dioxide disinfection process. For example, there are two CDIs presented for dichloroacetic acid; one CDI is based on an estimated environmental concentration of 0.081 μg/L resulting from the chlorination process, and one CDI is based on an estimated environmental concentration of 0.22 μg/L resulting from the chlorine dioxide process. Additionally, in some cases a chemical does not have a CDI; this is because no increase in concentration was measured as a result of the disinfection process.

Table 5-6. Chronic Daily Intake Results (mg/kg-day) by Exposure Pathway, for Noncancer Risk Calculation.

Residual or DBP - (Technology)	Water Ingestion	Inhalation	Dermal Contact
Chlorine w/o Reducing Agent - (chlorination)	7.0E-07	1.4E-04	0.25
Chlorine Dioxide w/o Reducing Agent - (chlorine dioxide)	1.2E-08	2.5E-06	0.0044
Chlorate - (chlorination)	7.4E-07	NA	0.27
Chlorate - (chlorine dioxide)	< 2.5E-06	NA	<0.89
Chlorite - (chlorine dioxide)	4.5E-06	NA	1.6
Chloroform - (chlorination)	7.4E-10	1.1E-07	2.4E-03
Monochloroacetic acid - (chorine dioxide)	1.3E-09	NA	5.8E-04
Dichloroacetic acid - (chlorination)	3.3E-09	NA	1.6E-03
Dichloroacetic acid - (chlorine dioxide)	9.0E-09	NA	4.4E-03
Trichloroacetic acid - (chlorination)	2.7E-09	NA	1.5E-03
Trichloroacetic acid - (chlorine dioxide)	2.7E-09	NA	1.5E-03

NA = not applicable

Table 5-7 Chronic Daily Intake Results (mg/kg-day) by Exposure Pathway, for Cancer Risk Calculation.

Residual or DBP - (Technology)	Water Ingestion	Inhalation	Dermal Contact
Chlorine w/o Reducing Agent - (chlorination)	3.0E-07	8.0E-05	0.11
Chlorine Dioxide w/o Reducing Agent - (chlorine dioxide)	5.3E-09	1.1E-06	0.0019
Chlorate - (chlorination)	<1.7E-07	NA	<0.061
Chlorate - (chlorine dioxide)	3.2E-07	NA	0.11
Chlorite - (chlorine dioxide)	1.9E-06	NA	0.7
Chloroform - (chlorination)	3.2E-10	4.7E-08	1.0E-03
Monochloroacetic acid - (chorine dioxide)	5.6E-10	NA	2.5E-04
Dichloroacetic acid - (chlorination)	1.4E-09	NA	7.0E-04
Dichloroacetic acid - (chlorine dioxide)	3.9E-09	NA	1.9E-03
Trichloroacetic acid - (chlorination)	1.4E-07	NA	8.1E-02
Trichloroacetic acid - (chlorine dioxide)	1.2E-09	NA	6.6E-04

NA = not applicable

5.3.4 Toxicity Assessment

The concepts and terminology associated with relevant toxicity endpoints for assessing risks to humans and aquatic organisms are briefly discussed below. A summary of the toxicity reference values available to characterize risks to both humans and aquatic organisms is provided in this chapter for the specific residuals and DPBs included in the disinfection pilot study. A more detailed review of the toxicity data available for these and other residuals and DPBs that may be associated with the disinfection approaches is presented in Chapter 8.0.

5.3.4.1 Human Health Effects-Noncancer

The reference dose (RfD) and reference concentration (RfC) are toxicity values used in risk assessments that provide quantitative information for noncancer effects (e.g., decreased body weight, reproductive toxicity, developmental toxicity, etc). The RfD is expressed in units of an intake rate of the chemical per unit body weight (mg/kg-day), while the RfC is expressed in units of concentration (mg/m^3). The use of the RfD and RfC in risk assessment is based on the assumption that thresholds exist for certain toxic effects (that is, at concentrations above the threshold an effect will likely be produced, and below that concentration it likely will not). The hazard quotient (HQ) method is simply the ratio of the expected dose divided by the appropriate toxicity reference value. Where possible, both the dose and the toxicity reference value should reflect the same exposure pathway (e.g., oral, dermal, inhalation). If the hazard quotient value is greater than one, then the chemical is classified as a COPC.

The general equation for the HQ, which is used to estimate risks to noncarcinogens, is given in the following equation.

$$HQ = \frac{CDI}{RfD} \quad or \quad \frac{C_{air}}{RfC}$$

where:

HQ	=	hazard quotient (unitless)
CDI	=	chronic daily intake (mg/kg-day)
C_{air}	=	concentration of the chemical in air (mg/m^3)
RfD	=	reference dose (oral, or equivalent dermal) (mg/kg-day)
RfC	=	reference concentration (inhalation) (mg/m^3)

5.3.4.2 Human Health Effects—Cancer

Carcinogenic risk assessments provide quantitative estimates of the upper bound increased risk of developing cancer following lifetime exposure via ingestion, inhalation, and/or dermal absorption (U.S. EPA, 1986). The toxicity reference value for cancer risk is referred to as either a cancer slope factor (CSF) or a unit risk. The CSF is analogous to the RfD in that unique chemical-specific CSF values may be developed for each route of exposure. Units for CSF are the inverse units of dose. The general equation for risk, which is used to assess the carcinogenic risk of a chemical, is given by the following equation.

$$Risk = CDI \times CSF$$

where:

Risk	=	cancer risk (unitless)
CDI	=	chronic daily intake (mg/kg-day)
CSF	=	cancer slope factor $(mg/kg\text{-}day)^{-1}$

5.3.4.3 Toxicity Reference Values for Human Health

Table 5-8 summarizes the toxicity reference values available for chlorine, chlorine dioxide, chlorate, chlorite, chloroform, monochloroacetic acid, dichloroacetic acid, and trichloroacetic acid. The oral RfD values for chlorine dioxide, chlorite, and chloroform are assumed to be applicable to both ingestion and dermal exposure pathways. This is a conservative assumption since it implies that bioavailability via both routes is equal (i.e., 100%) when, in fact, the skin can often serve as a protective barrier that reduces absorption. It also reflects an assumption that the skin is not a target organ of concern for noncancer health effects (e.g., skin irritation, etc.). Likewise, the oral CSF for chloroform is also assumed to be representative of the dermal CSF, with the same assumptions of equivalent bioavailability and no evidence of carcinogenicity associated with the skin. No inhalation toxicity reference values are available for this group of residuals and DPBs.

Table 5-8. Toxicity Reference Values Available for Chemicals with EECs (Source: IRIS).

Residual or DBP	CAS Registry Number	RfD[1] mg/kg-day	CSF[2] $(mg/kg\text{-}day)^{-1}$
Chlorine Residual	7782-50-5	1×10^{-1}	NA
Chlorine Dioxide Residual	10049-04-4	3×10^{-2}	NA
Chlorate	14866-68-3	NA	NA
Chlorite	7758-19-2	3×10^{-2}	NA
Chloroform	67-66-3	1×10^{-2}	1×10^{-2}
Monochloroacetic Acid	79-11-8	0.002	NA
Dichloroacetic Acid	79-43-6	1×10^{-3}	5×10^{-2}
Trichloroacetic Acid	76-03-9	NA	NA

[1] RfD was derived for oral exposure. A dermal RfD and inhalation RfC were not derived. Assuming equivalent bioavailability (100%) and that the primary noncancer health effect is not associated with the skin, the oral RfD is used to assess noncancer toxicity for the dermal pathway.
[2] CSF was derived for oral exposure. Unique CSF values for dermal and inhalation exposure were not derived. Assuming equivalent bioavailability (100%) and that the primary mode of action for carcinogenicity is not associated with the skin, the oral CSF is used to assess cancer risk for the dermal pathway.
NA = not available on IRIS

5.3.4.4 Ecological Effects—Definition of Measurement Endpoints

Ecological toxicity tests can measure lethal and/or sub-lethal effects. Acute toxicity tests are short-term tests (i.e., 96 hours or less) that measure the effects of exposure to relatively high concentrations of chemicals. The measurement endpoint generally reflects the extent of lethality. Chronic toxicity tests, on the other hand, generally are longer-term tests that measure the effects of exposure to relatively lower concentrations (U.S. EPA, 1997b). For a chronic toxicity test, the measurement endpoint concerns one or more sub-lethal effects (e.g., reproduction, growth) or both lethality and sub-lethal effects (U.S. EPA, 1992a). The duration of a chronic toxicity test depends upon the life-cycle of the test species.

Similar to noncancer human health effects, risk to aquatic organisms can be expressed as a hazard quotient (HQ). The HQ is the ratio of the EEC to the toxicity reference value. If the HQ is greater than one, then the chemical is classified as a COPC. For this disinfection demonstration project, risks to aquatic organisms were estimated using the HQ approach. Interpreting a "safe" or target level for the HQ involves risk-management considerations and goes beyond the scope of this literature review. For this analysis, HQ estimates for each chemical will be presented for purposes of characterizing relative risks among the potential chemicals.

5.3.5 Risk Characterization

5.3.5.1 Human Health

Noncancer hazard quotations and cancer risks were calculated using CDI results presented in Tables 5-6 and 5-7 and the corresponding toxicity reference values presented in Table 5-8. If toxicity reference values were not available for a particular pathway or chemical, then a calculation of risk or HQ could not be made (e.g., inhalation).

An example of risk and HQ calculations for dichloroacetic acid as a result of the chlorine dioxide disinfection process is given below.

Noncancer Hazard Quotient Calculation for Dichloroacetic Acid

Incidental Ingestion of Water

$$CDI = \frac{C_{water} \times CF_1 \times IR_{water} \times ET}{BW} \times \frac{EF}{365} \times \frac{ED}{AT}$$
$$= \frac{0.22 \times 0.001 \times 0.05 \times 3}{70} \times \frac{7}{365} \times \frac{30}{30}$$
$$= 9.0 \times 10^{-9}\ mg/kg - day$$

Inhalation

$$CDI = \frac{C_{air} \times IR_{air}}{BW} \times \frac{t_{event}}{24} \times \frac{EF}{365} \times \frac{ED}{AT}$$
$$= \frac{0 \times 2.0}{70} \times \frac{3}{24} \times \frac{7}{365} \times \frac{30}{30}$$
$$= 0\ mg/kg\ day$$

Dermal Contact

$$CDI = \frac{DA_{event} \times SA \times EV}{BW} \times \frac{EF}{365} \times \frac{ED}{AT}$$
$$= \frac{0.0009 \times 18000 \times 1}{70} \times \frac{7}{365} \times \frac{30}{30}$$
$$= 0.0044\ mg/kg\ day$$

The noncancer HQ associated with the cumulative CDI is, therefore:

$$HQ = \frac{\sum CDI}{RfD}$$
$$= \frac{(9.0E-9+0+0.0044)}{0.004}$$
$$= 1.1$$

Cancer Risk Calculation for Dichloroacetic Acid

Incidental Ingestion of Water

$$CDI = \frac{C_{water} \times CF_1 \times IR_{water} \times ET}{BW} \times \frac{EF}{365} \times \frac{ED}{AT}$$
$$= \frac{0.22 \times 0.001 \times 0.05 \times 3}{70} \times \frac{7}{365} \times \frac{30}{70}$$
$$= 3.9 \times 10^{-9}\ mg/kg-day$$

Inhalation

$$CDI = \frac{C_{air} \times IR_{air}}{BW} \times \frac{t_{event}}{24} \times \frac{EF}{365} \times \frac{ED}{AT}$$
$$= \frac{0 \times 2.0}{70} \times \frac{3}{24} \times \frac{7}{365} \times \frac{30}{70}$$
$$= 0\ mg/kg\ day$$

Dermal Contact

Since t_{event} (3 hours) > t* (1.3 hours), use

$$DA_{event} = FA \times K_p \times C_{water}\left[\frac{t_{event}}{1+B} + 2\tau_{event}\left(\frac{1+3B+3B^2}{(1+B)^2}\right)\right]$$
$$= 1 \times 0.001 \times 0.22\left[\frac{3}{1+0.0061} + 2 \times 0.5534\left(\frac{1+3 \times 0.0061 + 3 \times 0.0061^2}{(1+0.0061)^2}\right)\right]$$
$$= 0.0009\ mg/cm^2 event$$

and

$$CDI = \frac{DA_{event} \times SA \times EV}{BW} \times \frac{EF}{365} \times \frac{ED}{AT}$$
$$= \frac{0.0009 \times 18000 \times 1}{70} \times \frac{7}{365} \times \frac{30}{70}$$
$$= 0.0019\ mg/kg\ day$$

The cancer risk associated with the cumulative CDI is, therefore:

$$\begin{aligned} Risk &= CSF \times \sum CDI \\ &= (0.05) \times (3.9E-9+0+0.0019) \\ &= 9.5 \times 10^{-5} \end{aligned}$$

In addition to the noncancer risk HQ calculation performed for dichloroacetic acid, similar calculations were performed for the other chemicals presented in Table 5-6 if there was a published toxicity reference value (Table 5-8). Of these chemicals TRC from the chlorination disinfection process and chlorite and dichloroacetic acid from the chlorine dioxide disinfection process resulted in HQs greater than 1.0. This means that these chemicals would be identified as COPCs for the particular scenario defined in this screening-level risk assessment. However there are two important issues pertaining to this scenario that are vital to note: namely, the proper control of the disinfection processes and the conservative assumptions related to the dermal exposure pathway. Initially, the TRC concentrations used for this risk assessment were the result of the chlorination process prior to dechlorination, and the disinfection demonstration showed that proper dechlorination can eliminate TRC. Additionally, the high chlorite concentration in the chlorine dioxide effluent was residual chlorite in the chlorine dioxide product as a result of the generation method employed for this demonstration. Full-scale applications would employ generation processes that produce a chlorite-free product. Secondly, the scenario defined in this screening-level risk assessment for dermal exposure was very conservative due to the uncertainties associated with dermal exposure mechanics. As more is learned about the dermal exposures, the exposure assumptions can be refined, and it is likely that these chemicals will no longer be of concern.

Cancer risk calculations were performed for dichloroacetic acid, as presented above, and chloroform, the only two chemicals with published cancer slope factors (Table 5-8). Neither of these chemicals exceeded a risk of 1E-04, which is the trigger for a concern for cancer.

HQs for noncancer risk and cancer risk values are summarized in Table 5-9 for the residuals and DBPs resulting from both the chlorination and chlorine dioxide disinfection processes. Noncancer HQs greater than 1.0 and cancer risk values greater than 1E-04 are boldfaced.

Table 5-9. Toxicity Reference Values Available for Chemicals with EECs.

Residual or DBP (Technology)	CAS #	HQ[1]	Risk[2]
Chlorine w/o reducing agent (chlorination)	7782-50-5	**2.52**	NA
Chlorine dioxide w/o reducing agent (chlorine dioxide)	10049-04-4	0.15	NA
Chlorate (chlorination & chlorine dioxide)	14866-68-3	NA	NA
Chlorite (chlorine dioxide)	7758-19-2	**54**	NA
Chloroform (chlorination)	67-66-3	0.24	1.0E-05
Monochloroacetic acid (chlorination and chlorine dioxide)	79-11-8	0.29	NA
Dichloroacetic acid (chlorination)	79-43-6	0.41	9.5E-05
Dichloroacetic acid (chlorine dioxide)	79-43-6	**1.11**	3.5E-05
Trichloroacetic acid (chlorination and chlorine dioxide)	76-03-9	NA	NA

[1] HQ values exceeding 1 trigger a concern for noncancer health effects.
[2] Risk values exceeding 1E-04 trigger a concern for cancer.

As with the results of the screening-level assessment for cancer risk, assumptions associated with the estimate of dermal exposure, as well as the uncertainties associated with the use of an oral toxicity value to determine health effects from a dermal route, may need to be explored further in a more refined analysis. At a minimum, each of the chemicals for which HQ can be calculated would be identified as a COPC for the particular scenario defined in this report.

5.3.5.2 Risk Characterization-Aquatic Organisms

In terms of aquatic organism risk, the findings from this hazard identification (Chapter 8.0) were used to establish risk RBC as toxicity reference values. The reference concentrations for the chemicals were established as the lowest LC_{50} values found in the database, thus representing the most sensitive species. Table 5-9 summarizes the findings from the literature review.

Table 5-10 includes the estimated environmental concentrations for the residuals and DBPs expected in Onondaga Creek as well as the established toxicity reference value from the hazard identification. Comparing the expected environmental concentration to the reference value provides an HQ. If the HQ is greater than one, then the chemical should be classified as a COPC in context of this project, which means the chemical needs further investigation.

Table 5-10 is organized by disinfection technology type, namely chlorination and chlorine dioxide. If a value is not given for the expected environmental concentration for a given residual or DBP and a given disinfection technology, then an increase due to the disinfectant is not expected and therefore an HQ is not calculated.

Bioaccumulation was also considered as part of the screening-level risk assessment for all the measured disinfection residuals and DBPs (Table 4-2). All but one of these chemicals had bioaccumulation factors of less than 30 (i.e., threshold for "low" bioaccumulation potential). Only one DBP (2,4,6-trichlorophenol) was identified as having moderate potential for bioaccumulation, with a bioaccumulation factor of 55. This particular DBP was not measured in the disinfection demonstration because it was not considered a significant by-product.

5.3.6 Screening-level Risk Assessment Conclusions

Human health cancer risk calculations were calculated for dichloroacetic acid and chloroform, the only DBPs for which a cancer toxicity reference value was available. Regulatory thresholds for cancer risk may range from $1x10^{-6}$ to $1x10^{-4}$. For oral and inhalation pathways, cancer risks were orders of magnitude lower than this range, but for the dermal exposure pathway, the cancer risk was elevated but remained below the trigger value of $1x10^{-4}$. A number of conservative assumptions contribute to this calculation, including the use of an oral CSF to assess carcinogenicity via dermal contact and a series of exposure assumptions including: three hours of swimming, steady-state concentrations of these chemicals in the water during this period without losses due to vaporization, and dilution factors applied in the calculation of the EEC. However, even with these conservative assumptions the screening-level risk assessment results suggest that these chemicals do not warrant additional studies.

For human health noncancer HQ calculations, a common regulatory threshold HQ is 1.0. Of these chemicals, TRC from the chlorination disinfection process and chlorite and dichloroacetic acid from the chlorine dioxide disinfection process resulted in HQs greater than 1.0 and were therefore identified as COPCs for the particular scenario defined in this screening-level risk assessment. For these three chemicals, HQ values were well below 1.0 for both ingestion and inhalation but well above 1.0 for dermal contact. The same uncertainties apply for HQ as were discussed above for cancer risk. This suggests that additional study may be warranted to develop more refined estimates of both exposure and toxicity.

Based on the screening-level risk assessment approach for aquatic organisms, the chlorination disinfection technology produced one COPC, namely residual chlorine. The residual chlorine concentration used in the assessment was representative of the chlorination process without dechlorination. The COPC classification only means the chemical may need further study. In the case of residual chlorine, it is well documented that concentrations exceeding 500 μg/L can cause toxicity to aquatic organisms; therefore it is routinely required that chlorinated effluents be treated with a quenching agent to reduce the concentration of residual chlorine.

Based on the screening-level risk assessment approach for aquatic organisms, the chlorine dioxide disinfection technology produced two COPC, namely residual chlorine and chlorite. The COPC classification only means the chemical may need further study. The chlorine and chlorite were by-products of the particular chlorine dioxide production method; this could be avoided by using a generation method that produces a chlorine- and chlorite-free product, which would be more appropriate for a full-scale application. If, however, these by-products are created in the generation method, they can be eliminated by the use of quenching agents, such sodium bisulfite and ferrous sulfate.

Table 5-10. Hazard Identification Literature Review Findings.

Chemical	CAS NO	Acute Toxicity Value			Chronic Toxicity Value		
		Species	Value (µg/L)	Endpoint	Species	Value (µg/L)	Endpoint
Hypochlorous acid	7790-92-3	FW Fish: FW Invert: SW Fish: SW Invert:	100 ND ND ND	2-hr LC50 ND ND ND	ND		
Chlorine dioxide	10049-04-4	FW Fish: FW Invert: SW Fish: SW Invert:	6400000 350 ND 500000	96-hr LC50 96-hr LC50 ND 48-hr LC50	ND		
Chlorite ion	7758-19-2	FW Fish: FW Invert: SW Fish: SW Invert:	100000 10 75000 580	96-hr LC50 48-hr EC50 96-hr LC50 96-hr LC50	FW Fish: FW Invert: SW Fish: SW Invert:	50600 ND ND ND	312-hr LC50 ND ND ND
Chlorate ion (chlorate hydrate)	14866-68-3	ND			ND		
Chloroform	67-66-3	FW Fish: FW Invert: SW Fish: SW Invert:	13300 2000 ND 2400	96-hr LC50 1-hr LC50 ND 1-hr LC50	FW Fish: FW Fish: FW Invert: FW Invert: SW Fish: SW Invert:	1240 122000 235000 6300 ND ND	672-hr LC50 240-hr NOEC 216-hr LC50 504-hr NOEC ND ND
Monochloroacteic acid (metabolite of DCA)	79-11-8	FW Fish: FW Invert: SW Fish: SW Invert:	ND 77000 ND ND	ND 48-hr EC50 ND ND	FW Fish: FW Invert: SW Fish: SW Invert:	ND 32000 ND ND	ND 504-hr NOEC ND ND
Dichloroacetic acid (DCA)	79-43-6	FW Fish: FW Invert: SW Fish: SW Invert:	ND ND ND 23000	ND ND ND 96-hr LC50	ND		
Trichloroacetic acid	76-03-9	FW Fish: FW Invert: SW Fish: SW Invert:	1050000 1200 9300000 4800000	96-hr LC50 24-hr LC50 96-hr LC50 96-hr LC50	ND		

= Circled value is the lowest available ecotoxicity value and is therefore used as the screening toxicity value for the risk assessment.

Table 5-11. Hazard Quotient for Chlorination and Chlorine Dioxide Residuals and By-products.

	Chlorination			Chlorine Dioxide		
Chemical	Expected Environmental Concentration Value (µg/L)	Minmum LC_{50} Value (µg/L)	Hazard Quotient (Conc/LC_{50})	Expected Environmental Concentration Value (µg/L)	Minmum LC_{50} Value (µg/L)	Hazard Quotient (Conc/LC_{50})
Hypochlorous acid (TRC)	880	100	**9**	200	100	**2**
Chlorine dioxide				2	350	<0.01
Chlorate ion	500	3,186,000	<0.01	920	3,186,000	<0.01
Chlorite ion				6	10	**0.6**
Chloroform	1	2,000	<0.01			
Monochloroacteic acid				1.7	77,000	<0.01
Dichloroacetic acid	4.2	23,000	<0.01	12	23,000	<0.01
Trichloroacetic acid	3.4	1,200	<0.01	3.5	1,200	<0.01

5.4 Operations

Operation of chemical disinfectant systems should ideally provide variable disinfectant dosing during wet weather events to match the variability of the discharge. Such variability can be addressed in either one of two ways or a combination of 1) program control based on a history of the discharge, and 2) real-time control. The short duration of first-flush periods may warrant program control proportioned to flow. The period following the first-flush of the storm may warrant real-time control such as TRC in addition to flow to avoid overdosing.

Operation of UV disinfection for wet weather flows may follow more of a program control. Because most often collection systems will have short duration events in the order of hours, UV disinfection equipment will likely be operated at a constant lamp intensity. Optimizing disinfection efficiency would then be limited to the adjustment of the wet weather flow rate (instead of intensity) to ensure equal dose exposure to the flow.

Development of a program control should include a review of historical water quality data at a discharge and using the data to predict disinfectant dose requirements. By establishing a relationship between water quality and disinfection dose, a dose curve or feed system algorithm can be developed for the expected water quality range. This is discussed in further detail below.

5.4.1 Developing a Disinfection Dose Feed Program

A disinfection dose feed program can be an algorithm that includes flow and total suspended solids (TSS) as variables. The feed program should be flexible to allow for fine tuning based on operational experience. Such a program can optimize the disinfection dose and minimize residuals.

The general process of developing a feed program algorithm should include developing a relationship between TSS concentration and wet weather flow rate as well as a relationship between TSS concentration and disinfectant dose. Once these relationships are established, a simple algorithm can be developed that relates the wet weather flow rate to a disinfectant dose. This algorithm can be used during design to size components and also provide for operational control. This is the approach implemented with successful results for the Columbus Water Works CSO facilities in Columbus, GA (Boner and Turner, 2002).

As an illustration of the above approach, a typical CSO discharge from Onondaga County was used to develop a feed program algorithm. Development of the algorithm was based on actual TSS and flow rate measurements from a typical CSO discharge as well as the dose-response curves from the disinfection demonstration (Chapter 4.0).

A plot of flow rates versus TSS measurements for a typical CSO discharge in Onondaga County is exhibited in Figure 5-2. The power regression curve is plotted through these data that represent the outer envelope of TSS occurrence during wet weather flow events at this discharge. At a peak flow rate of 175 MGD the TSS concentration is expected to be approximately 50 mg/L and at a flow rate of 115 MGD the expected TSS is approximately 130 mg/L.

A normalized dose-response curve for the chlorination process is exhibited in Figure 5-3. This dose-response curve is similar to those presented in Chapter 4.0; however, the dose is normalized by TSS instead of COD. In order to use this dose-response curve to develop the feed program algorithm, a targeted effluent (i.e., end-of-pipe) fecal coliform concentration must be defined. In the case of Onondaga County the targeted end-of-pipe fecal coliform concentration is 2,000 CFU/100 mL. This fecal coliform concentration is the y value on the dose-response curve, and therefore by using the equation of the line, an x value of 0.0073 is calculated. The x value of 0.0073 is the targeted chlorine dose normalized by TSS and temperature.

The relationships developed in Figures 5-2 and 5-3 can be used together to develop an algorithm that results in a required chlorine dose range based on the wet weather flow rate. This algorithm is presented in Figure 5-4 for chlorine. Using this algorithm, a targeted chlorine dose range of approximately 20 mg/L to 7 mg/L is calculated for the CSO discharge. It is important to note that the average chlorine dose of 9 mg/L presented in Section 5.2.2 falls on the lower end of this feed program algorithm because smaller storms with lower flow rates are more frequent and thus represent more the average storm condition.

Following these same procedures, algorithms for chlorine dioxide, ozonation, and UV were calculated and are presented in Figures 5-5, 5-6, and 5-7.

The practitioner must understand that the feed system (for chemical) and the flow control system (for UV disinfection) need to operate for the entire dose range in order to have the ability to achieve the effluent bacterial criteria. Out in the collection system, the range of dosing may be more pronounced and possibly with sharper peaks as compared to a centralized location due to dilution effects. The practitioner should understand that the algorithm control should include adjustment factors that the operator can fine-tune as performance experience is gained. The practitioner should also understand that disinfectant dose and reducing agent dose should be slightly greater than what is required to ensure meeting the effluent bacterial criteria and residual target, respectively.

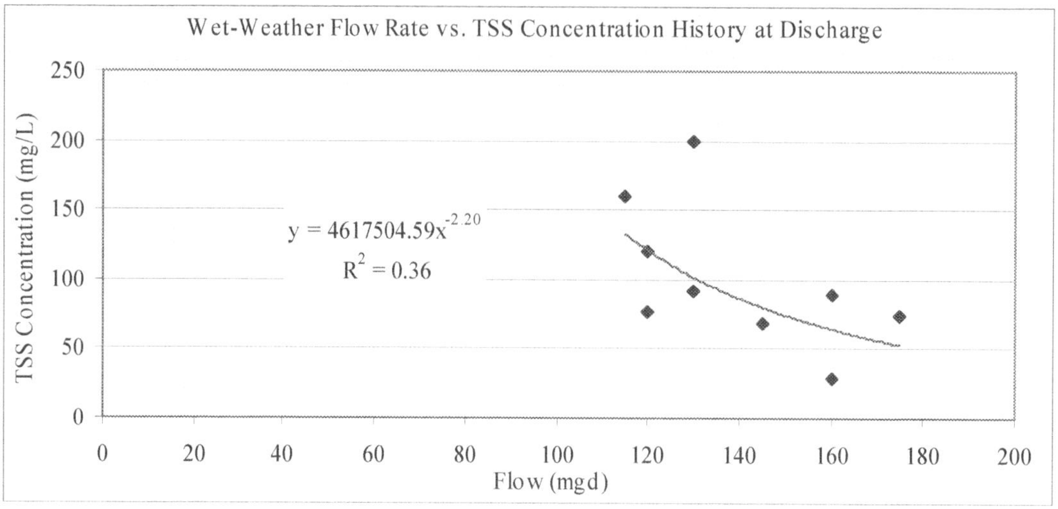

Figure 5-2. Wet Weather Flow Rate versus TSS Concentration, Site Specific Discharge.

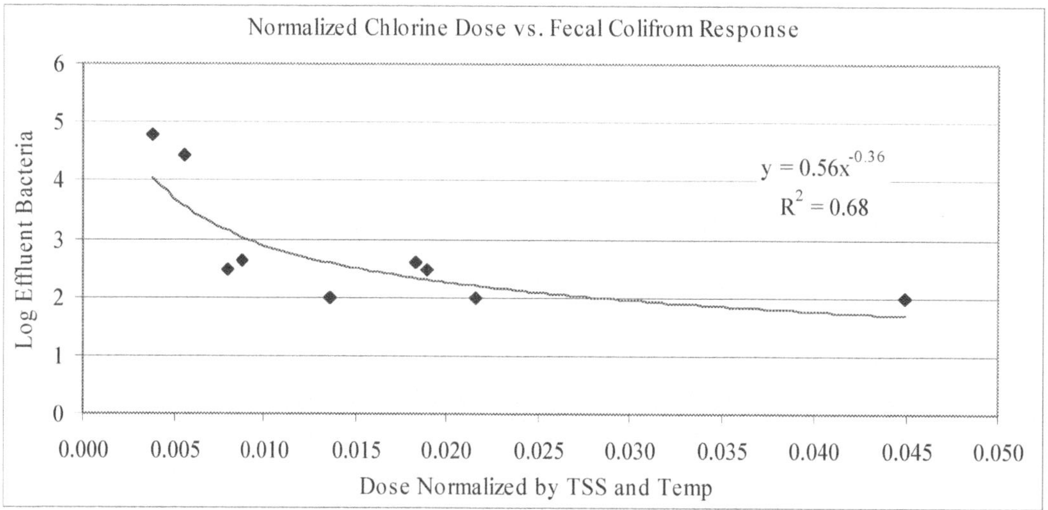

Figure 5-3. Chlorine Dose-Response Curves Normalized by TSS and Temperature.

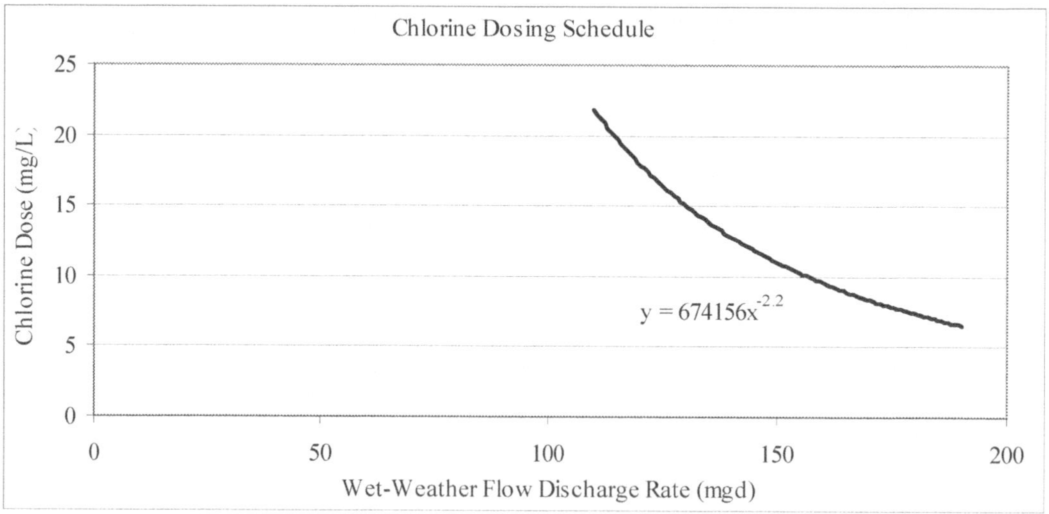

Figure 5-4. Chlorine Dosing Schedule for Site Specific Discharge.

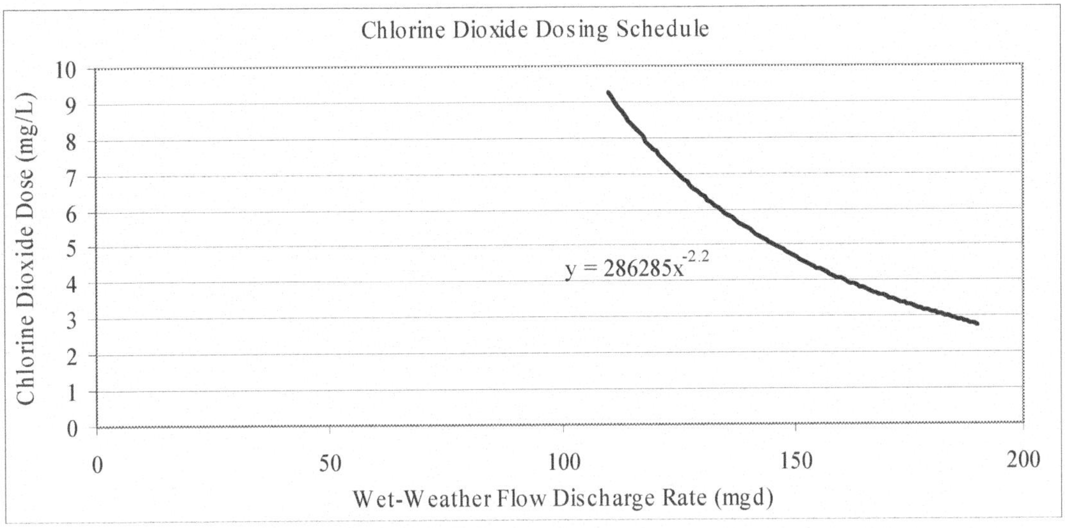

Figure 5-5. Chlorine Dioxide Dosing Schedule for Site Specific Discharge.

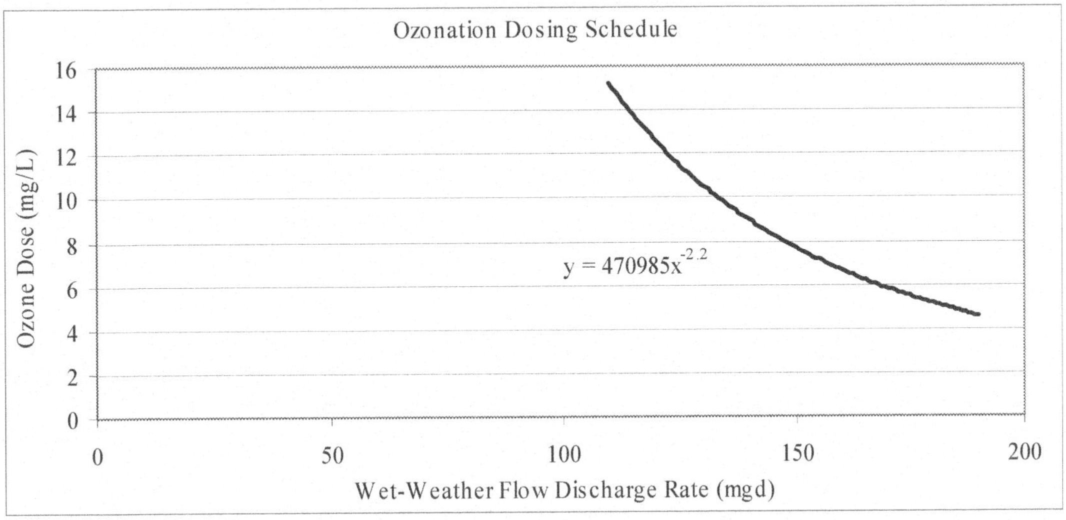

Figure 5-6. Ozone Dosing Schedule for Site Specific Discharge.

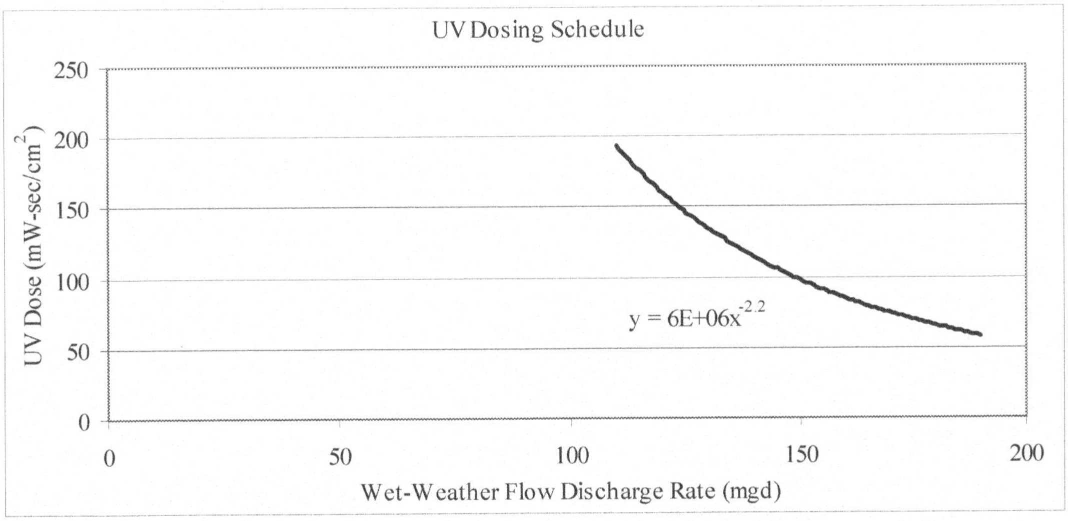

Figure 5-7. UV Dosing Schedule for Site Specific Discharge.

5.5 Conceptual Costs

A conceptual cost analysis was performed on the four disinfection technologies. The purpose of the analysis was to provide a relative cost comparison of the technologies over a wide range of design flow rates. The costs represent equipment required for disinfection including contact tanks, chemical storage and feed equipment, electrical and instrumentation and building services, as appropriate. The costs do not include ancillary structures such as process piping to and from the disinfection equipment or wet weather flow pumping.

The conceptual costs for the different technologies were based on the average doses presented in Section 5.2.2 and also the peak dose determined from the dose feed program exercise in Section 5.4.1.

Three different peak flow rates were used in the analysis, namely 1 MGD, 10 MGD, and 100 MGD. The assumptions made for developing the conceptual costs for each disinfection technology are presented below. Both capital costs and chemical and power costs were developed for each technology at each of the three flow rates, except where otherwise noted. Facility structures, such as the contact chamber and chemical storage tanks, were sized based on a peak wet weather flow rate and/or an assumed wet weather storm event flow volume. Chemical and power costs were based on an assumed annual wet weather flow volume treated. The storm event volumes and the annual volumes for the three different flow rates are presented below. Also included with this information is the assumed duration of flow through the facilities for an event as well as an annual basis.

5.5.1 Storm Event and Annual Volumes

As stated above, conceptual costs were developed for peak flow rates of 1 MGD, 10 MGD, and 100 MGD. In addition to peak flow rates, storm volumes and annual volumes were required to size facilities and develop capital costs and chemical and power costs. Storm volumes were established by developing synthetic hydrographs for the specified peak flows, assuming a storm duration of four hours. Annual volumes were based on the assumption that 12 storm events of four-hour duration occur each year.

Storm Event Volume

- 1-MGD Storm Event Volume: 87,500 gallons
- 10-MGD Storm Event Volume: 875,000 gallons
- 100-MGD Storm Event Volume: 8,750,000 gallons
- Note: Event Duration for 1, 10, and 100 MGD Storms = 4 hours

Annual Volume

- 1-MGD Annual Volume: 1.1 million gallons
- 10-MGD Annual Volume: 10.5 million gallons
- 100-MGD Annual Volume: 105 million gallons
- Note: Total Annual Event Durations for 1, 10, and 100 MGD Storms = 48 hours (12 events @ 4 hours each)

A present worth (year 2003, ENR= 6850[1]) was developed for each technology at each flow rate. The present worth analysis assumed a 10-year design life at a 7% interest rate.

5.5.2 Chlorination/Dechlorination

Table 5-11 presents the major components of chlorination/dechlorination facilities and their associated costs. The assumptions used for facility sizing and cost estimating are listed below.

5.5.2.1 General Assumptions

- Maximum Dose: 20 mg/L
- Average Dose: 9 mg/L
- Strength of Chlorine (Purchased): 12%
- Strength of Chlorine (Stored after Degradation): 7.5%
- Strength of Sodium Bisulfite (Dechlorination Agent): 38%

5.5.2.2 Capital Cost Assumptions

- Disinfection Contact Chamber: Chamber sized for 6 minutes of retention(5 minutes for chlorination and 1 minute for dechlorination)
- Chlorination and Dechlorination Mixers and Controls: Mixers sized based on ½ hp per MGD of wet weather flow (NSF International and U.S. EPA, 2002)
- Chlorination and Dechlorination Metering Pumps and Controls: Metering pumps sized based on the following mass balance approach:

  Qw * Cw = Qmp * Cp

  Where: Qw is the wet weather flow

  Cw is the maximum dose

  Qmp is the metering pump flow rate

  Cp is the concentration of the degraded chlorine

- Chlorination and Dechlorination Storage Tanks: Tanks sized based on storing sufficient quantity of degraded chemical to disinfect two back-to-back storm event volumes at the average dose

5.5.2.3 Chemical and Energy Costs Assumptions

- Liquid Chlorine and Sodium Bisulfite: Chemical quantities based on average dose and annual volume of wet weather flow treated
- Mixer Power: Power consumption based on mixer horsepower (chlorination and dechlorination mixers) and the annual wet weather flow duration (48 hours)

Table 5-12. Chlorination/Dechlorination Conceptual Capital Costs and Chemical and Energy Costs.

[1] *Engineering News Record* Construction Cost Index

1 MGD	Quantity		Cost Estimate
Capital Costs			
Disinfection Contact Chamber	4,150 gallons	$	11,983
Chlorination Mixer(s) and Control	1-1/2 hp Mixers and Controls	$	15,600
Dechlorination Mixer(s) and Control	1-1/2 hp Mixers and Controls	$	15,600
Chlorination Metering Pumps and Controls	0.2 gpm	$	2,400
Dechlorination Metering Pumps and Controls	0.05 gpm	$	1,200
Chlorination Storage Tanks	1-50 gallon Tank	$	150
Dechlorination Storage Tanks	1-10 gallon Tank	$	100
Misc. Piping and Valves	Misc. Piping and Valves	$	2,000
Sensor/Telemetry System	Sensor/Telemetry System	$	10,000
Storage Building with HVAC	8' x 8' Pre-fabricated Storage Unit	$	10,000
	Disinfection Facility Cost	**$**	**69,033**
Chemical and Energy Costs			
Liquid Sodium Hypochlorite	150 gallons	$	68
Sodium Bisulfite	25 gallons	$	28
Mixer Power	45 kwhr	$	4
	Annual Chemical and Energy Costs	**$**	**99**

10 MGD	Quantity		Cost Estimate
Capital Costs			
Disinfection Contact Chamber	41,500 gallons	$	117,549
Chlorination Mixer(s) and Control	1-5 hp Mixers and Controls	$	22,425
Dechlorination Mixer(s) and Control	1-5 hp Mixers and Controls	$	22,425
Chlorination Metering Pumps and Controls	2 gpm	$	9,200
Dechlorination Metering Pumps and Controls	0.5 gpm	$	2,640
Chlorination Storage Tanks	1-300 gallon Tank	$	1,725
Dechlorination Storage Tanks	1-50 gallon Tank	$	150
Misc. Piping and Valves	Misc. Piping and Valves	$	10,000
Sensor/Telemetry System	Sensor/Telemetry System	$	18,000
Storage Building with HVAC	100 square feet	$	13,950
	Disinfection Facility Cost	**$**	**218,064**
Chemical and Energy Costs			
Liquid Sodium Hypochlorite	1,500 gallons	$	675
Sodium Bisulfite	250 gallons	$	275
Mixer Power	450 kwhr	$	36
	Annual Chemical and Energy Costs	**$**	**986**

100 MGD	Quantity		Cost Estimate
Capital Costs			
Disinfection Contact Chamber	415,000 gallons	$	1,175,488
Chlorination Mixer(s) and Control	3-20 hp Mixers and Controls	$	99,180
Dechlorination Mixer(s) and Control	3-20 hp Mixers and Controls	$	99,180
Chlorination Metering Pumps and Controls	20 gpm	$	28,500
Dechlorination Metering Pumps and Controls	5 gpm	$	12,540
Chlorination Storage Tanks	2-1,500 gallon Tanks	$	13,200
Dechlorination Storage Tanks	2-200 gallon Tanks	$	2,300
Misc. Piping and Valves	Misc. Piping and Valves	$	60,000
Sensor/Telemetry System	Sensor/Telemetry System	$	25,000
Storage Building with HVAC	400 square feet	$	55,800
	Disinfection Facility Cost	**$**	**1,571,188**
Chemical and Energy Costs			
Liquid Sodium Hypochlorite	15,000 gallons	$	6,750
Sodium Bisulfite	2,500 gallons	$	2,750
Mixer Power	4,500 kwhr	$	360
	Annual Chemical and Energy Costs	**$**	**9,860**

5.5.3 Chlorine Dioxide

Table 5-12 presents the major components of chlorine dioxide facilities and their associated costs. The assumptions used for both sizing and cost estimating are presented below.

5.5.3.1 General Assumptions

- Maximum Dose: 15 mg/L
- Average Dose: 5 mg/L
- Chlorine dioxide generated on-site and on-demand with a system that uses chlorine gas to oxidize sodium chlorite

5.5.3.2 Capital Costs Assumptions

- Disinfection Contact Chamber: Chamber sized for five minutes of retention (disinfection only, no chlorite or chlorine dioxide reduction required)
- Chlorine Dioxide Generator: Generator sized to meet peak wet weather flow rate.
- Chlorine Dioxide Mixer(s) and Controls: Mixers sized based on ½ hp per MGD of wet weather flow (NSF International and U.S. EPA, 2002)
- Chlorine Dioxide Metering Pumps and Controls: Same mass balance equation as Chlorination Metering Pumps and Controls ($Qw * Cw = Qmp * Cp$)

5.5.3.3 Chemical and Energy Costs Assumptions

- Gaseous Chlorine: Volume based on producing sufficient quantity of chlorine dioxide for an average dose and annual volume of wet weather flow treated.
- Sodium Chlorite: Same as Gaseous Chlorine
- Mixer Power: Power consumption based on mixer horsepower (chlorine dioxide only) and the annual wet weather flow duration (48 hours)

Table 5-13. Chlorine Dioxide Conceptual Capital Costs and Chemical and Energy Costs.

	Quantity	Cost Estimate
1 MGD		
Capital Costs		
Disinfection Contact Chamber	4,150 gallons	$ 11,983
Chlorine Dioxide Mixer(s) and Control	1-1/2 hp Mixers and Controls	$ 15,600
Chlorine Dioxide Metering Pumps and		$ 25,000
Chlorine Dioxide Generator		$ 75,000
Chlorine Gas Storage	10-pound gas cylinder	$ -
Sodium Chlorite Storage	1-10 gallon Tank	$ 100
Misc. Piping and Valves	Misc. Piping and Valves	$ 5,000
Sensor/Telemetry System	Sensor/Telemetry System	$ 20,000
Storage Building with HVAC	100 square feet	$ 13,950
	Disinfection Facility Cost	**$ 166,633**
Chemical and Energy Costs		
Gaseous Chlorine	30 lbs	$ 21
Preserved Sodium Chlorite	30 lbs	$ 216
Mixer Power	23 kwhr	$ 2
Generator Power	17 kwhr	$ 1
	Annual Chemical and Energy Costs	**$ 239**
10 MGD		
Capital Costs		
Disinfection Contact Chamber	41,500 gallons	$ 117,549
Chlorine Dioxide Mixer(s) and Control	1-5 hp Mixers and Controls	$ 22,425
Chlorine Dioxide Metering Pumps and		$ 50,000
Chlorine Dioxide Generator		$ 110,000
Chlorine Gas Storage	50-pound gas cylinder	$ -
Sodium Chlorite Storage	1-100 gallon Tank	$ 250
Misc. Piping and Valves	Misc. Piping and Valves	$ 15,000
Sensor/Telemetry System	Sensor/Telemetry System	$ 32,000
Storage Building with HVAC	300 square feet	$ 41,850
	Disinfection Facility Cost	**$ 389,074**
Chemical and Energy Costs		
Gaseous Chlorine	300 lbs	$ 184
Preserved Sodium Chlorite	300 lbs	$ 2,160
Mixer Power	225 kwhr	$ 18
Generator Power	170 kwhr	$ 14
	Annual Chemical and Energy Costs	**$ 2,362**
100 MGD		
Capital Costs		
Disinfection Contact Chamber	347,000 gallons	$ 982,878
Chlorine Dioxide Mixer(s) and Control	3-20 hp Mixers and Controls	$ 99,180
Chlorine Dioxide Metering Pumps and		$ 100,000
Chlorine Dioxide Generator		$ 400,000
Chlorine Gas Storage	1/4-ton gas cylinder	$ -
Sodium Chlorite Storage	1-1,000 gallon Tank	$ 4,400
Misc. Piping and Valves	Misc. Piping and Valves	$ 80,000
Sensor/Telemetry System	Sensor/Telemetry System	$ 50,000
Storage Building with HVAC	1000 square feet	$ 139,500
	Disinfection Facility Cost	**$ 1,855,958**
Chemical and Energy Costs		
Gaseous Chlorine	3,000 lbs	$ 1,575
Preserved Sodium Chlorite	3,000 lbs	$ 21,600
Mixer Power	2,250 kwhr	$ 180
Generator Power	6,750 kwhr	$ 540
	Annual Chemical and Energy Costs	**$ 23,355**

5.5.4 Ozonation

Table 5-13 presents the major components of ozonation facilities and their associated costs. The assumptions used in the analysis are presented below.

5.5.4.1 General Assumptions

- Maximum Dose: 24 mg/L
- Average Dose: 15 mg/L
- Ozone generated on-site and on-demand with a system that uses stored liquid oxygen

5.5.4.2 Capital Costs Assumptions

- Ozonation system comes complete with:
    - Oxygen Storage
    - Ozone Generator
    - Eductor and Contactor
    - Sensors and Controls
- Ozonation system only requires telemetry equipment for sensors and a control building

5.5.4.3 Chemical and Energy Costs Assumptions

- Liquid Oxygen: Volume based on producing enough ozone for an average dose and annual volume of wet weather flow treated

Power: Power consumption based on producing enough ozone for an average dose and annual volume of wet weather flow treated from industrial-grade liquid oxygen

Table 5-14. Ozonation Conceptual Capital Costs and Chemical and Energy Costs.

1 MGD	Quantity		Cost Estimate
Capital Costs			
Complete Ozone System Including:	Ozone System	$	300,000
Oxygen Storage			
Ozone Generator			
Eductor and Contactor			
Sensors and Controls			
Telemetry System	Telemetry System	$	7,500
Storage Building with HVAC	100 square feet	$	13,950
	Disinfection Facility Cost	**$**	**321,450**
Chemical and Energy Costs			
Liquid Oxygen	270 lbs	$	2,310
Ozone Generator Power	160 kwhr	$	13
	Annual Chemical and Energy Costs	**$**	**2,323**
10 MGD	**Quantity**		**Cost Estimate**
Capital Costs			
Complete Ozone System Including:	Ozone System	$	1,000,000
Oxygen Storage			
Ozone Generator			
Eductor and Contactor			
Sensors and Controls			
Telemetry System	Telemetry System	$	7,500
Storage Building with HVAC	100 square feet	$	13,950
	Disinfection Facility Cost	**$**	**1,021,450**
Chemical and Energy Costs			
Liquid Oxygen	2,700 lbs	$	23,100
Ozone Generator Power	1,600 kwhr	$	128
	Annual Chemical and Energy Costs	**$**	**23,228**

Note: Ozone is not recommended at this flow rate due to lack of full-scale facilities at this size and cost.

Note 2: Because ozanation was not cost-competitive at the 10 MGD flow rate, a cost estimate for 100 MGD was not performed.

5.5.5 Ultraviolet Light

Table 5-14 presents the major components of UV facilities and their associated costs. The assumptions used in the analysis are presented below.

5.5.5.1 General Assumptions

- Design Dose: 100 mWsec/cm2
- UV unit chamber type with medium pressure, high intensity bulbs

5.5.5.2 Capital Costs Assumptions

- UV system come complete with:
  - Chamber
  - Controls
  - AutoWipe
  - UV Monitor
- UV system only requires telemetry equipment for sensors and a control building

5.5.5.3 Chemical and Energy Costs Assumptions

- No chemicals required
- Power: Power consumption based on producing the design UV dose and annual volume of wet weather flow treated

Table 5-15. Ultraviolet Light Conceptual Capital Costs and Chemical and Energy Costs.

1 MGD			
	Quantity		Cost Estimate
Capital Costs			
Complete UV System Including:	UV System	$	60,000
Chamber			
Controls			
AutoWipe			
UV Monitor			
Hydraulic Controls	Flow Distribution Equipment	$	25,000
Telemetry System	Telemetry System	$	7,500
Storage Building with HVAC	100 square feet	$	13,950
	Disinfection Facility Cost	**$**	**106,450**
Chemical and Energy Costs			
UV Power	1,800 kwhr	$	144
	Energy Costs	**$**	**144**
10 MGD			
	Quantity		Cost Estimate
Capital Costs			
Complete UV System Including:	UV System	$	500,000
Chamber			
Controls			
AutoWipe			
UV Monitor			
Hydraulic Controls	Flow Distribution Equipment	$	75,000
Telemetry System	Telemetry System	$	7,500
Storage Building with HVAC	100 square feet	$	13,950
	Disinfection Facility Cost	**$**	**596,450**
Chemical and Energy Costs			
UV Power	18,000 kwhr	$	1,440
	Energy Costs	**$**	**1,440**
100 MGD			
	Quantity		Cost Estimate
Capital Costs			
Complete UV System Including:	UV System	$	4,400,000
Chamber			
Controls			
AutoWipe			
UV Monitor			
Hydraulic Controls	Flow Distribution Equipment	$	300,000
Telemetry System	Telemetry System	$	7,500
Storage Building with HVAC	200 square feet	$	27,900
	Disinfection Facility Cost	**$**	**4,735,400**
Chemical and Energy Costs			
UV Power	180,000 kwhr	$	14,400
	Energy Costs	**$**	**14,400**

5.5.6 Conceptual Cost Summary

Chlorination/dechlorination and chlorine dioxide are cost competitive at each of the three flow rates, namely 1 MGD, 10 MGD, and 100 MGD. It is likely, however, that the cost of maintaining a chlorine dioxide generator, which was not included in these conceptual costs, may give a cost advantage to chlorination/dechlorination. It is also likely that at flow rates greater than 100 MGD that the cost of purchasing a larger chlorine dioxide generator to meet peak demands may become cost prohibitive compared to increasing chemical storage for chlorination/dechlorination.

UV appears to be cost competitive at 1 MGD and loses advantages at 10 and 100 MGD. However, for this conceptual cost analysis, UV should be considered cost competitive to chlorination/dechlorination and chlorine dioxide at the 10 MGD flow rate. At a flow rate of 100 MGD, UV has lost its cost competitiveness as compared to chlorination/dechlorination and chlorine dioxide. This is most likely due the large UV reactors that are required for the peak flow that have no redundancy value at low flows. In fact many reactors may sit unused at the facility but once per year when the peak design flow rate occurs. In the case of chlorination/dechlorination and chlorine dioxide, the same equipment is used at low flows and high flows and therefore is a better use of capital. This comparison is solely based on this conceptual cost analysis and does not consider other factors, such as ease of operation, DBPs, toxicity, and public perception. When considering other factors such as these, UV may become more competitive to chlorination/dechlorination and chlorine dioxide at these higher flow rates.

Ozonation was slightly more costly than the other technologies at the 1 MGD flow rate and not competitive at the 10 MGD flow rate. Because ozonation was not competitive at 10 MGD, a cost estimate for 100 MGD was not performed. The high cost of ozonation can be attributed to the cost of generation, the complexity of injected ozone into the wet weather flow and the cost of producing ozone. Even the most efficient ozone generators require proportionally much greater electrical costs than the other technologies.

A summary of the conceptual costs for these technologies is presented in Figure 5-8.

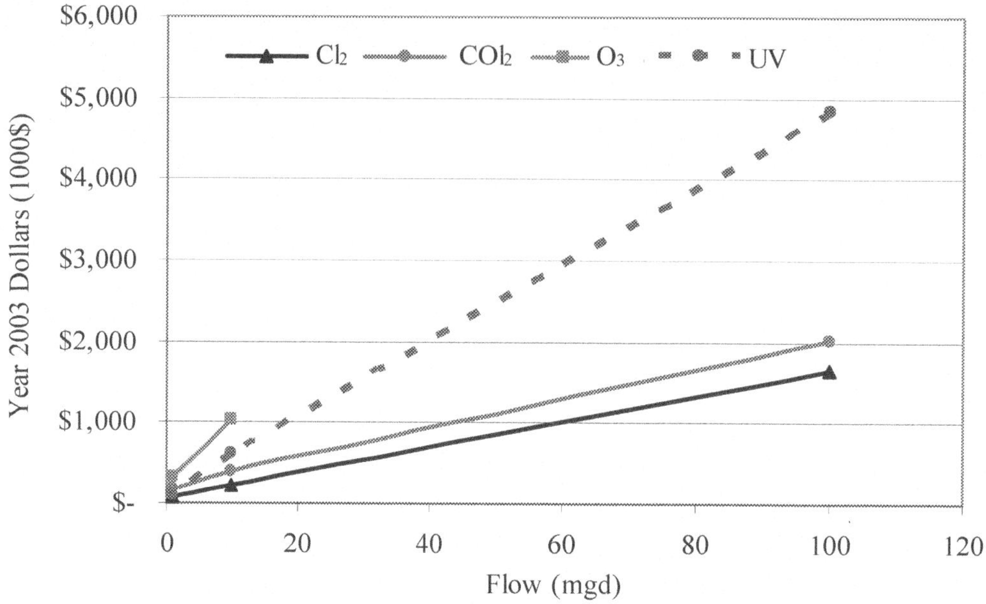

Figure 5-8. Conceptual Capital Costs and Chemical and Energy Costs.

CHAPTER 6.0

COMMUNICATION

6.1 Decision-Making Framework Communication Plan

The public in general has played a significant role in deciding the level and type of pollution abatement as codified in the Clean Water Act Amendment of 1972. Public participation was a required element of facility planning in the construction grant process. Most recently, public involvement for CSO abatement has been identified in the U.S. EPA guidance, *Coordinating CSO Long Term Planning with Water Quality Standards Reviews*. This guidance requires public involvement in the development of the long-term control plan, which often involves disinfection.

Involving the public in the decision-making framework for disinfecting wet weather flow requires a communication plan. A communication plan is generally intended to be an umbrella, guiding all of the municipality's external communications. However, more concise communication plans should be proactively developed for specific and appropriate projects, such as the design and construction of wet weather flow disinfection facilities. Considering that disinfecting wet weather flows will include some type of benefit-risk analysis, a communication plan that has elements of risk communication is important. Consequently the communication plan for this project included both elements of effectively communicating information to the public and responding to the public reaction to the information.

During this project, an important component of the decision-making framework, as described in Chapter 2.0, was implementing a communication plan through stakeholder involvement along the way as depicted in Figure 2-1. The local stakeholders were involved throughout the project by means of monthly progress reports and workshops. Not included in the decision-making framework, but equally important, was public outreach. The public was involved through a public workshop.

The communication plan implemented in this project is common for typical wastewater treatment programs today. Four meetings with the stakeholders were held in the form of workshops. The stakeholders were involved at the beginning of the project to ensure that all information relevant to the assessment was identified (*Site-Specific and General Information Input*) ; twice in the middle of the project to ensure that assumptions regarding the wet weather flow discharges and receiving water were valid, to plan for collecting data through a disinfection demonstration, and to review the demonstration findings (*Benefit/Risk/Feasibility Assessment*); and at the end of the project to assist in making risk-management decisions and exclude disinfection technologies that do not meet criteria (*Technology Selection*).

Working in collaboration with the Onondaga County Department of Water Environment Protection, the local stakeholders included informed citizens and regulatory personnel as listed in Chapter 2.0. The local stakeholders represented the public at the "decision-making table" throughout the process. The criteria for stakeholder selection were:

- Representativeness of a group or entity with a defined concern;
- Knowledge of a defined concern; and
- Availability and willingness to participate on a prescribed schedule.

In addition to the stakeholder workshop, the general public was invited to participate in a public workshop where local concerns and perceptions were voiced regarding the tradeoffs between microbial risks to people versus chemical risks to people and to the aquatic ecosystem. Concerns brought forth by the public included: location of wet weather flow treatment facilities in a residential areas, odors from the facility, safe handling of disinfectants at the facility, carcinogenic effects of DBPs, and aquatic impacts of DBPs.

6.2 Communication Tools

The type and form of communication tools selected for any effort should be tailored to the audience they are intended to reach. In this project, the targeted audience included the local stakeholders and the concerned public as well as WERF subscribers by way of this report. The communication objectives were to:

- Include Onondaga County and stakeholders in the decision-making process; and
- Communicate to the concerned public the benefits and risks of disinfecting wet weather flows.

The following four tools were created to communicate and report the findings of this project.

1. A public workshop for the purpose of communicating the benefits and risks of disinfecting wet weather flows. The presentation slides can be found on the project webpage (communication tool #2)
2. A webpage that summarizes the information presented at the public workshop. Questions and answers from the public workshop are also presented on the webpage. This web page is hosted by Onondaga County and can be accessed at: http://www.lake.onondaga.ny.us/ol301.htm
3. A detailed final report to include all results and analyses of the disinfection demonstration study
4. An executive summary that is an abridged version of the detailed report tailored to the interests of managers and decision makers

6.3 Communicating Risk

Communicating risks to a community involves presenting information in an unbiased fashion and in terms that are understandable to the community. As important as effective communication of risks is being aware of public perceptions, which most notably transpires in the form of public outrage over a particular situation. When it comes to communicating risks, it is important to recognize that outrage is as important as the risk itself. "When people are outraged, they tend to think the risk is more serious than it is. If you make a list of environmental health risks in order of how many people die from such risks and then make another list of risks in order of the extent of alarm to the general public, the two lists will be very different" (Sandman, 1993).

Hazard and outrage can be categorized into the following four groups as described by Sandman (2003). Based on findings from this project, wet weather flow disinfection can be related to the second and third groups "Stakeholder Relations" and "Outrage Management," as discussed in the following paragraphs.

1. *Public Relations: High Hazard, Low Outrage*

Since the hazard is high, or at least higher than the outrage, the task is often to provoke more outrage.

This paradigm is called public relation (PR) because the essence of PR is talking to people who are not interested. But this is also the domain of health education, safety training, and environmental activism. Urging people to take some risk more seriously is the kind of risk communication industrial hygienists do most often.

2. *Stakeholder Relations: Moderate Hazard, Moderate Outrage*

Stakeholders are relatively more attentive than the public—neither too apathetic nor too upset to listen. The objective of stakeholder relations is to discuss the issues openly and rationally, explaining your views and responding to audience questions and concerns. Instead of the mass media, stakeholder relations rely on interpersonal dialogue, supplemented by specialized media such as newsletters and web sites. Technical details may need to be explained as this audience may really wants to hear them.

3. *Outrage Management: Low Hazard, High Outrage*

Thisoccurs when the audience is outraged, largely at you. And while the outrage may be justified for some non-technical reasons (for example, you may have been less than courteous), it isn't technically justified; the hazard is low. A core group of "fanatics" is usually accompanied by a larger, less outraged constituency watching to see how the controversy evolves. The goal is to reduce audience outrage by listening and acknowledging. The controversy ends and the goal is achieved when the "fanatics" declare victory or their constituency thinks they have won enough.

In outrage management, the medium is face-to-face dialogue, though this time the "audience" does most of the talking. Barriers include the audience's outrage at you, your own outrage at the audience, coming to grips with the need to focus on outrage when you'd really rather talk about substance, and occasionally the presence of the media.

4. *Crisis Communication: High Hazard, High Outrage*

This is a relatively rare but enormously important sort of risk communication. The audience is huge and very upset. The outrage is even greater than in outrage management. It's also different—more fear and misery than anger. The task, therefore, is to help the audience bear its fear and misery. Key strategies include avoiding over-reassurance, sharing dilemmas, being human and empathic, providing things to do, and acknowledging uncertainty. A good example of this type of communication was New York City's Mayor Rudolph Giuliani's handling of the September 11, 2001, terrorist attacks on New York City.

6.4 Lessons Learned

Selecting local stakeholders to participate in the decision-making process was an effective approach in soliciting local values to shape in the decisions regarding the selection of appropriate disinfection technologies, as well as the design and construction of wet weather flow disinfection facilities. It was important that these stakeholders were involved early in the process and that they represented a wide range of views from the community. These stakeholders responded well to monthly progress reports and semi-annual workshops. The lunchtime workshops were held during the business week and scheduled for one and a half-hours with lunch provided. Slide shows were presented at each of the workshops with time for questions and answers. These workshops provided a comfortable atmosphere where stakeholders were willing to share ideas and concerns regarding the progress of the project.

Even though the local stakeholders represented a broad range of views, none of the stakeholders were entirely opposed to the idea of disinfecting wet weather flows.

The public workshop presented an opportunity for communicating preliminary findings of this project, including the risks of disinfection, to the public. Risk communication techniques were employed. In summary, the workshop was successful. Many good questions were asked along with statements of concerns and issues. It was clear that some attendees, who were completely opposed to the idea of treating wet weather flows, including disinfection, had as much interest in voicing their opinions as learning about wet weather flow disinfection. These people were from neighbor organizations initially favoring the concept to store wet weather flows in underground storage tanks so that there would be minimal visual structures and the flows could be pumped to the WWTP for treatment with little or no discharge to the receiving water. In general, it appeared that the attendees gained a better understanding of the risks and benefits of disinfection from this workshop, and the few people in complete opposition represented a vocal minority.

Major concerns brought forth by the public at the workshop included: location of wet weather flow treatment facilities in residential neighborhoods, odors from the facility, safe handling of disinfectants at the facility, carcinogenic effects of DBPs, and environmental effects of DBPs.

Although the risks associated with wet weather flow disinfection were shown to be low, based on reactions from the workshop, the select public opposition was high. The communication plan that was prepared at the onset of the project included "Stakeholder Relations" but did not include "Outrage Management." An outrage management technique could have included both supporters and critics as stakeholders throughout the decision-making process with the intent of gaining trust along the way.

CHAPTER 7.0

CONCLUSIONS

- Chlorination, chlorine dioxide, ozonation, and UV technologies used for disinfection of wet weather flow discharges can all achieve U.S. EPA guideline criteria for bacteria (U.S. EPA 2002).
- Normalizing dose by water quality constituents provides a reliable base for developing dose-response relationships.
- Temperature, COD, and TSS proved to be the most effective normalization parameters; however, COD/temperature normalization provides the best basis for comparing disinfection technologies.
- TSS normalization as it relates to flow rate can provide the best means of real-time control at a wet weather flow disinfection facility.
- At the targeted doses, chlorination (18 mg/L) and chlorine dioxide (7mg/L) produce disinfection residuals and DBPs.
- At the targeted doses, ozonation (23 mg/L), and UV (100 mWsec/cm^2) do not produce disinfections residuals and DBPs.
- Typical methods of chlorination produce TRC, chlorate THMs, and HAAs.
- Using the chlorite/acid method to generate chlorine dioxide produces TRC, chlorate, chlorite, and HAAs.
- Concentrations of residuals and DBPs as measured for chlorination and chlorine dioxide in this demonstration, with the exception of chloroform for the chlorination process, were found to represent an acceptable (i.e., low) risk to human health as defined by the U.S. EPA (1989).
- TRC concentrations from the chlorination process and TRC and chlorite concentrations from the chlorine dioxide process, as measured in this demonstration, represent an unacceptable risk to aquatic organisms as defined by the U.S. EPA (1992a).

- Reducing agents can control TRC in the chlorination process and TRC and chlorite in the chlorine dioxide process, providing an acceptable risk to aquatic organisms as defined by the U.S. EPA (1992a).

- Commercially available full-scale chlorine dioxide generation methods can produce TRC and chlorite-free product that provide acceptable risk to aquatic organisms as defined by the U.S. EPA (1992a).

- Chlorination and chlorine dioxide disinfection facilities are comparable and can be most cost effective up to flows of 100 MGD when compared to UV disinfection and ozonation.

- UV disinfection using medium pressure, high intensity lamps can be cost effective up to 10 MGD. Ozonation can be cost effective up to 1 MGD. Ozonation is the most expensive technology due to operational complexity in combination with high concentration-time product (Ct) values.

- Final selection of a disinfection technology may not be based solely on cost but by the result of operational and/or public preference.

- Although the risks of disinfecting wet weather flow discharges are low, concerns and opposition by the public may be high resulting in the need for "outrage management"..

- Involvement of stakeholders and the public throughout the decision-making process can be a method to reduce opposition. Preparing a communications plan at the onset of the project is a critical step.

- Public involvement can be managed by communication, involvement and gaining trust along the way.

CHAPTER 8.0

HAZARD IDENTIFICATION LITERATURE REVIEW

8.1 Hazard Identification Summary

This chapter summarizes the relevant literature and toxicity values for disinfection residuals and DBPs measured in the effluent of each treatment technology: chlorine disinfection with dechlorination, chlorine dioxide disinfection, ozonation disinfection, and UV disinfection.

8.2 Hazard Identification Databases

The majority of the toxicity literature relevant to human health was obtained from the following databases: Hazardous Substances Data Bank (HSDB), International Agency for Research on Cancer (IARC), and IRIS. The majority of the data relevant to aquatic organisms was obtained from Aquatic Toxicity Information Retrieval (AQUIRE). These databases are briefly summarized below.

8.2.1 HSDB—Hazardous Substances Data Bank

HSDB is a toxicology data file on the National Library of Medicine's (NLM) Toxicology Data Network (TOXNET). It focuses on the toxicology of potentially hazardous chemicals. It is enhanced with information on human exposure, industrial hygiene, emergency handling procedures, environmental fate, regulatory requirements, and related areas. All data are referenced and derived from a core set of books, government documents, technical reports, and selected primary journal literature. The web address is http://toxnet.nlm.nih.gov/cgi-bin/sis/htmlgen?HSDB.

8.2.2 IARC Monographs—International Agency for Research on Cancer

In 1969, the International Agency for Research on Cancer (IARC) initiated a program to evaluate the carcinogenic risk of chemicals to humans and to produce monographs on individual chemicals. These monographs are widely considered to be authoritative, independent assessments of the carcinogenic risks posed to humans by a variety of agents, mixtures, and exposures. The web address is http://193.51.164.11/cgi/iHound/Chem/iH_Chem_Frames.html.

8.2.3 IRIS—Integrated Risk Information System

IRIS, prepared and maintained by U.S. EPA, is an electronic database containing information on human health effects that may result from exposure to various chemicals in the environment. IRIS was initially developed for EPA staff in response to a growing demand for consistent information on chemical substances for use in risk assessments, decision-making and regulatory activities. Health assessment information on a chemical substance is included in IRIS only after a comprehensive review of chronic toxicity data by U.S. EPA health scientists from several program offices and the Office of Research and Development. Background information and explanations of the methods used to derive the values given in IRIS are provided in the background documents available from the IRIS web site. Regulatory toxicity values are presented along with descriptive and quantitative summaries of the available study data. Toxicity values that represent the chronic noncarcinogenic health effects include oral reference doses (RfD, mg/kg-day) and inhalation reference concentrations (RfC, m3/kg/day). Toxicity values that represent the carcinogenic health effects include oral slope factors (SFo, [mg/kg-day]-1), inhalation slope factors (SFi, [m3/kg-day]-1), and oral and inhalation unit risks. Toxicity values and supporting information are posted on IRIS periodically, and occasionally values that are posted on IRIS are reviewed to ensure that they represent the most current information pertaining to that chemical. As a result, it is important for users of the toxicity information on IRIS to check the database to verify that the information reflects the current EPA tox value. The web address for this database is http://www.epa.gov/ngispgm3/iris/index.html.

8.2.4 ECOTOX/AQUIRE

The ECOTOX (an EPA ecotoxicity database) system represents an integration of AQUIRE, PHYTOTOX (an EPA terrestrial plant toxicity database), and TERRETOX (an EPA wildlife toxicity database), which are three existing EPA databases that contain ecotoxicity information for aquatic life, terrestrial plants, and wildlife, respectively.

The AQUIRE database was established in 1981 by U.S. EPA, Mid-Continent Ecology Division, Duluth, MN (MED-Duluth). Scientific papers published both nationally and internationally on the toxic effects of chemicals to aquatic organisms and plants are collected and reviewed for AQUIRE. Independently compiled laboratory data files that include AQUIRE parameters and meet the quality assurance criteria are also included. Aquatic literature has been reviewed and entered into the system, the majority of which encompasses test results reported from 1970 to the present. The web address for this database is http://www.epa.gov/ecotox/.

8.3 Chlorine Residuals

In pure water, chlorine forms elemental chlorine (Cl_2), chloride ion (Cl^-), and hypochlorous acid (HOCl). As pH increases, hypochlorous acid dissociates to hypochlorite ion (OCl^-). The term "free chlorine" (free available chlorine, free residual chlorine) refers to the concentrations of elemental chlorine, hypochlorous acid, and hypochlorite ion that collectively occur in water. Several factors, including chlorine concentration, pH, temperature, exposure to light and the presence of catalysts or organic material, affect the stability of free chlorine in aqueous solution (U.S. EPA, 1992b as cited by U.S. EPA, 2004b). An update literature search conducted since the IRIS posting identified only one article that is relevant to this discussion: Kallen and Robert (2000).

8.3.1 Mammalian (Human) Toxicity

Data on the toxicity or carcinogenicity of hypochlorous acid and the hypochlorite ion in humans are inadequate, though a number of animal studies exist. The EPA (U.S. EPA, 2004b) has derived an RfD of $1x10^{-1}$ mg/kg-day based on a chronic study in rats in which no adverse effects occurred (NTP, 1992). No data were available concerning the carcinogenicity of these chemicals from EPA, IARC, or ACGIH (U.S. EPA, 2004b; IARC, 2001; ACGIH, 2000).

In a two-year study, rats and mice (70/sex/group) were administered chlorinated drinking water containing 0, 70, 140, or 275 ppm of available chlorine (NTP, 1992). According to the authors, these concentrations correspond to doses of 4.2, 7.3, and 13.6 mg/kg-day for male rats; 4.2, 7.8, and 14.4 mg/kg-day for female rats; 7.4, 14.0, and 24 mg/kg-day for male mice; and 7.6, 14.2, and 24.2 mg/kg-day for female mice. All animals were subjected to complete necropsy and a histopathologic examination at completion of the study. Results from these evaluations were unremarkable for both rats and mice: survival, hematological parameters, body weight and histological parameters among treated rats and mice were similar to controls. Only a slight, apparently dose-related decrease in water consumption was noted in both rats and mice. The highest dose administered was identified as the no observed adverse effect level (NOAEL): 13.6 or 14.4 mg/kg-day for male and female rats, respectively. The 14.4 mg chlorine/kg in female rats was selected as the basis of the RfD (U.S. EPA, 2004b).

An updated literature search, conducted since the IRIS posting, identified only one article that is relevant to this discussion. Kallen and Robert (2000) surveyed the Swedish Medical Birth Registry and official data on drinking water chlorination to study the health effects of consuming drinking water that has been disinfected with chlorine dioxide or sodium hypochlorite. For their study, three cohorts were identified and compared: women who lived in areas where drinking water was disinfected with chlorine dioxide, women who lived in areas that used sodium hypochlorite disinfection, and women who lived in areas where there was no chlorination of the drinking water. The authors reported a statistically significant increase in the number of gestational and developmental parameters, including short gestational duration, low birth weight, short body length, and very small head circumference in areas using sodium hypochlorite. These effects were not statistically significant in areas using chlorine dioxide disinfection. Because the exposure information in this study was limited to the chlorination method used and did not determine the amount of disinfection residuals or by-products in the drinking water, the study does little to add to the database regarding the association between adverse effects and exposure to disinfection residuals or by-products.

8.3.1.1 Noncancer Effects

The NOAEL from the National Toxicology Program (NTP) study (which is the basis for the RfD) is supported by NOAELs from other chronic and subchronic rat studies of additional toxicologic parameters, including hematology, serum chemistry, and histopathology (U.S. EPA, 2004b). Developmental and reproductive toxicity of chlorine have been examined in mice and rats but with suboptimal studies. Due to the chemical relationship between chlorine and monochloramine, reproductive and developmental studies for monochloramine may be used to satisfy data gaps for chlorine (U.S. EPA, 2004b).

8.3.1.2 Carcinogenic Effects

There are no data available on the carcinogenic potential of chlorine, hypochlorous acid, or hypochlorite ion (U.S. EPA, 2004b). A large number of epidemiology studies have reported an association between the use of chlorine disinfection and increased cancer risk (NAS, 1987). However, only the study by Cantor et al. (1987) was designed properly. This study suggested an association between chlorination of surface drinking water and bladder cancer. However, neither this study, nor the others, was sufficiently robust to provide any indication of causality or to be generalized to all chlorinated waters (Bull and Kopfler, 1991).

8.3.2 Aquatic Ecotoxicity

For hypochlorous acid, a total of 22 unique records were found in AQUIRE. Two records were excluded from the analysis due to insufficient information, e.g., no study endpoint. The remaining 20 records were grouped as freshwater fish. The database contained information on three species of freshwater fish.

The most sensitive acute freshwater effects were seen in a fish, the emerald shiner (*Notropis atherinoides*). The two-hour study reported an LC_{50} for mortality of 100 ppb. None of the records contained information on freshwater invertebrates. None of the records contained information on chronic freshwater endpoints. None of the records contained information on saltwater endpoints.

8.4 Chlorine Dioxide, Chlorite, and Chlorate

The toxicities of chlorine dioxide, chlorite, and chlorate are described together, since chlorine dioxide is fairly unstable and rapidly dissociates, predominantly into chlorite, chlorate, and chloride. There is interconversion among these species in water (before administration to animals) and in the gut (after ingestion) (U.S. EPA 1994a as cited in U.S. EPA, 2000d). Therefore, what exists in water or the stomach of a test animal is a mixture of these chemicals (i.e., chlorine dioxide, chlorite, chlorate) and possibly their reaction products with the gastrointestinal contents. Each of these chemicals is rapidly absorbed from the gastrointestinal tract and is widely distributed throughout the body as chloride (U.S. EPA, 2000d). For this reason, these chemicals were the subjects of an IRIS Toxicological Review (U.S. EPA, 2000d), wherein they were assessed together. An updated literature search,conducted since the IRIS toxicological review was posted,only identified one relevant article (Kallen and Robert, 2000), so the information below presents a review of that paper and a summary of the IRIS Toxicological Review for chlorine dioxide, chlorite, chlorate (U.S. EPA, 2000d).

8.4.1 Mammalian (Human) Toxicity

Although there are a number of human reports and animal studies in the literature regarding exposures to these chemicals, neither the noncancer effects nor the cancer effects have been adequately assessed (U.S. EPA, 2000d). The animal toxicity databases for chlorine dioxide, chlorite, and chlorate are fairly comprehensive, composed of subchronic and chronic studies, reproductive and developmental studies, as well as toxicokinetic and mechanistic information (U.S. EPA, 2000d). Multiple animal studies have consistently shown adverse neurodevelopmental effects, such as brain weight and behavioral measures (U.S. EPA, 2000d). The oral RfD for chlorine dioxide, chlorite, or chlorate is 3×10^{-2} mg/kg-day (U.S. EPA, 2004b).

Kallen and Robert (2000) surveyed the Swedish Medical Birth Registry and official data on drinking water chlorination to study the health effects of consuming drinking water that has been disinfected with chlorine dioxide or sodium hypochlorite. For their study, three cohorts were identified and compared: women who lived in areas where drinking water was disinfected with chlorine dioxide, women who lived in areas that used sodium hypochlorite disinfection, and women who lived in areas where there was no disinfection of the drinking water. The authors reported a statistically significant increase in a number of gestational and developmental parameters, including short gestational duration, low birth weight, short body length, and very small head circumference, in areas using sodium hypochlorite. These effects were not statistically significant in areas using chlorine dioxide disinfection. Because the exposure information in this study was limited to the chlorination method used and did not determine the amount of disinfection residuals or by-products in the drinking water, the study does little to add to the database regarding the association between adverse effects and exposure to disinfection residuals or by-products resulting from chlorine dioxide.

No human studies assessing the carcinogenic potential of chlorine dioxide, chlorite, or chlorate were located. Chlorine dioxide carcinogenicity has not been tested in animal bioassays. Chlorite was not shown to increase tumor incidences in rats and mice; these studies are considered inadequate for assessing human carcinogenicity (U.S. EPA, 2000d). Studies of chlorate by EPA and National Institute of Environmental Health Sciences (NIEHS) showed thyroid follicular cell hyperplasia in rats with exposures as short as 21 days. It was proposed that chlorate may interfere with thyroid uptake of iodine. Thus, a quantitative carcinogenicity assessment cannot be performed for these chemicals.

8.4.1.1 Noncancer Effects

The toxicity resulting from oral exposure to chlorine dioxide has not been adequately assessed. A number of studies have consistently found developmental effects following *in utero* exposure or postnatal gavage (experimentally placing the chemical in the stomach) administration of 14 mg/kg-day chlorine dioxide. The effects include altered brain development and decreased locomotor or exploratory activity (U.S. EPA, 2000d). These data suggest that the critical effect of chlorine dioxide is neurodevelopmental toxicity (U.S. EPA, 2000d). Other endpoints have been studied, but results are inconsistent (U.S. EPA, 2000d). A number of studies have also examined the toxicity of chlorite and chlorate. However, as with chlorine dioxide, data suggest that developmental toxicity consistently appears to be the most sensitive effect of oral chlorite or chlorate exposure (U.S. EPA, 2000d). Thus, for chlorine dioxide, chlorite, and chlorate, the most sensitive effect following oral exposure is neurodevelopmental delay (U.S. EPA, 2000d).

8.4.1.2 Carcinogenic Effects

Under the current guidelines for cancer assessment (U.S. EPA, 1986), chlorine dioxide is classified as Group D, not classifiable as to human carcinogenicity because of inadequate data in humans and animals (U.S. EPA, 2004b). The update literature search identified no new human or animal studies assessing the carcinogenic potential of chlorine dioxide. In general, human studies have not found adverse effects in individuals consuming low concentrations (0.04–0.15 mg/kg-day) of chlorine dioxide or chlorite in experimental studies or consuming drinking water disinfected with chlorine dioxide (U.S. EPA, 2000d).

The carcinogenic potential of concentrates prepared from drinking water that had been treated with chlorine dioxide was tested in mice by Miller (1986 as cited in U.S. EPA, 2000d). The concentrates did not increase incidence of lung adenomas in Strain A mice, skin tumor frequency in mice, or incidence of gamma-glutamyl transpeptidase positive foci (a measure of preneoplastic changes) in rat livers. Robinson (1986 as cited in U.S. EPA, 2000d) found significant increases in skin thickness in SENCAR mice immersed in chlorine dioxide, suggesting that high concentrations of chlorine dioxide are capable of inducing hyperplastic responses in the mouse skin. Both positive and negative results have been found in genotoxicity studies of chlorine dioxide (U.S. EPA, 2000d).

Under the current guidelines (U.S. EPA, 1986), chlorite and chlorate are classified as Group D, not classifiable as to human carcinogenicity because of inadequate data in humans and animals (U.S. EPA, 2004b). The update literature search revealed no new human or animal studies assessing the carcinogenic potential of chlorite. Chlorite was tested for potential carcinogenicity in drinking water studies conducted in rodents (rats by Kurokawa, Takayama, Konishi, 1986; mice by Yokose, Uchida, Nakae, 1987 as cited in U.S. EPA, 2000d). These studies do not provide sufficient evidence to draw conclusions as to the carcinogenic potential of chlorite in humans (U.S. EPA, 2000d). In the rat study (Kurokawa, Takayama, Konishi, 1986 as cited in U.S. EPA, 2000d), exposure to sodium chlorite did not significantly increase the incidence of tumors. The short exposure duration (85 weeks) and high incidence of Sendai viral infection in control and exposed rats limit the use of this study to assess carcinogenicity (U.S. EPA, 2000d).

In the mouse drinking water study of chlorite (Kurokawa, Takayama, Konishi,, 1986; Yokose, Uchida, Nakae, 1987 as cited in U.S. EPA, 2000d), significant increases in liver and lung tumors were observed in male mice. Tumor incidences observed in the test animals were within the range of values of historical controls in the study laboratory and in the NTP laboratories (Kurokawa, Takayama, Konishi, 1986 as cited in U.S. EPA, 2000d). Chlorite has been shown to be mutagenic in *in vitro* assays for reverse mutations and chromosome aberrations and in an *in vivo* assay of micronucleus induction in mice, although other *in vivo* assays were negative (U.S. EPA, 2000d).

8.4.2 Aquatic Ecotoxicity

For chlorine dioxide, a total of 34 unique records were found in AQUIRE. Fourteen records were excluded from the analysis due to insufficient information, e.g., no study endpoint. The remaining 20 records were grouped as follows: eight as freshwater fish, eight as freshwater invertebrates, and four as saltwater invertebrates. The database contained information on three species of freshwater fish, three freshwater invertebrates, and four saltwater invertebrates. A total of three unique records were found in HSDB. All three records were grouped as freshwater fish.

The most sensitive acute freshwater effects were seen in a fish, the fathead minnow juvenile (*Pimephales promelas*). The 96-hour study reported an LC_{50} for mortality of 20 ppb. Freshwater fish are less sensitive than freshwater invertebrates. The most sensitive freshwater fish was red rasbora harlequinfish (*Rasbora heteromorpha*). The 96-hour study reported an LC_{50} for mortality of 6,400,000 ppb. The most sensitive chronic effects were seen in a freshwater fish, the rainbow/Donaldson trout (*Oncorhynchus mykiss*). The 312-hour study (13 days) reports an LC_{50} for mortality of 50,600 ppb.

The most sensitive acute saltwater effects were seen in invertebrates, green or European shore crabs (*Carcinus maenas*), cockles (*Cerastoderma edule*), common shrimp/sand shrimp (*Crangon crangon*), and Aesop shrimp (*Pandalus montagui*). Each of the 48-hour studies reported LC_{50} values for mortality of 500,000 ppb. None of the records contained sufficient information on saltwater invertebrates.

None of the records contained information on chronic freshwater or saltwater endpoints.

For sodium chlorite, a total of 22 unique records were found in AQUIRE. Four records were excluded from the analysis due to insufficient or incompatible information (e.g., no study endpoint). The remaining 18 records were grouped as follows: 12 as freshwater fish; four as freshwater invertebrates; one as saltwater fish; and one as saltwater invertebrates. The database contained information on two species of freshwater fish, one freshwater invertebrate, one saltwater fish, and one saltwater invertebrate.

The most sensitive acute freshwater effects were seen in an invertebrate, the water flea (*Daphnia magna*). The 48-hour study reported an EC_{50} of 10 ppb. Freshwater fish are less sensitive than freshwater invertebrates. The most sensitive freshwater fish were rainbow trout/donaldson trout (*Oncorhynchus mykiss*) and bluegills (*Lepomis macrochirus*). Each of the 96-hour studies reported LC_{50} values for mortality of 100,000 ppb. The most sensitive chronic freshwater effects were seen in a fish, the rainbow trout/Donaldson trout. The 312-hour study reported an LC_{50} for mortality of 50,600 ppb.

The most sensitive acute saltwater effects were seen in an invertebrate, the opossum shrimp (*Americamysis bahia*). The 96-hour study reported an LC_{50} of 580 ppb. Saltwater fish are less sensitive than saltwater invertebrates. The most sensitive saltwater fish was the sheepshead minnow (*Cyprinodon variegates*). The 96-hour study reported an LC_{50} for mortality of 75,000 ppb. None of the records contained information on chronic saltwater endpoints.

No unique records for chlorate or sodium chlorate were found in the AQUIRE database. However sodium chlorate is considered nontoxic to fish. The possible 48-hour LC_{50} for various species of fish was 10,000 mg/L (The Royal Society of Chemistry 1983). A study of pulp and paper waste treated with chlorine dioxide reported an LC_{50} for *Daphnia magna* Straus, using the 48-hour acute lethality test, of at 3,162 mg/L chlorate (Dosdall et al., 1997).

8.5 THM: Chloroform

8.5.1 Mammalian Toxicity

Based on a chronic study in beagle dogs (Heywood et al., 1979), EPA has derived an RfD of $1x10^{-2}$ mg/kg-day. EPA also considers this reference dose ($1x10^{-2}$ mg/kg-day) to be protective against cancer effects because the data suggest that the first step in carcinogenesis is cytotoxicity and cell regeneration (U.S. EPA, 2004b), and these effects are included in the noncancer RfD. In 1999, IARC classified chloroform as possibly carcinogenic to humans, based on insufficient evidence of human carcinogenicity and sufficient evidence of carcinogenicity in animals (IARC, 2001). ACGIH (2000) has classified chloroform as category A3: confirmed animal carcinogen with unknown relevance to humans.

8.5.1.1 Noncancer Effects

Heywood et al. (1979) exposed beagle dogs to chloroform in a toothpaste base (0.5 mL of toothpaste base/kg/day) in gelatin capsules. A control group composed of 16 males and 16 females received the vehicle, and additional control groups of eight animals/sex were administered an alternative toothpaste or were left untreated. Experimental groups of eight male and eight female dogs received 15 or 30 mg chloroform/kg/day for 6 days/week. Treatment was continued for 7.5 years. Fatty cysts, considered to be treatment-related, were observed in livers of some dogs in both treatment groups. Nodules of altered hepatocytes were considered treatment-related but not dose-dependent. A dose-related increase in serum enzymes that indicate injury to the liver or heart was noted in the high-dose animals. The lowest observed adverse effect level (LOAEL) was determined to be 12.9 mg/kg-day, and an RfD was set at 0.01 mg/kg-day (U.S. EPA, 2004b).

8.5.1.2 Carcinogenic Effects

Chloroform has been tested for carcinogenicity in eight strains of mice, two strains of rats and in beagle dogs. Jorgenson et al. (1985) administered chloroform (pesticide quality and distilled) in drinking water to male rats and female mice at concentrations corresponding to 19, 38, 81, and 160 mg/kg-day for rats and 34, 65, 130, and 263 mg/kg-day for mice. A significant dose-related increase in renal tumors in rats was observed in the highest dose group. The liver tumor incidence in female mice was not significantly increased. This study was specifically designed to measure the effects of low doses of chloroform. In support of these data, a 78-week gavage bioassay by the National Cancer Institute (NCI, 1976) in rats and mice reported that a significant increase in kidney epithelial tumors was observed in male rats, as were highly significant increases in hepatocellular carcinomas in mice of both sexes. Also, liver cancer was observed in mice gavaged with chloroform (Eschenbrenner and Miller, 1945; Rudali, 1967).

The majority of tests for genotoxicity of chloroform have been negative. The carcinogenicity of chloroform may be a function of its metabolism to phosgene, which is known to cross-link DNA. A host-mediated assay using mice indicated that chloroform was metabolized *in vivo* to a form mutagenic to bacteria. Likewise urine extracts from chloroform-treated mice were mutagenic (U.S. EPA, 2004b). Based on the lack of genotoxic action, the weight of evidence suggests that chloroform is carcinogenic only after inducing cell death and regeneration. Hence, the RfD for chloroform, which accounts for these effects is protective against cancer.

8.5.2 Aquatic Ecotoxicity

A total of 231 unique records were found in AQUIRE. Ninety-five records were excluded from the analysis due to insufficient or incompatible information, e.g., no study endpoint. The remaining 136 records were grouped as follows: 103 as freshwater fish, 25 as freshwater invertebrates, and eight as saltwater invertebrates. The database contained information on 10 species of freshwater fish, six freshwater invertebrates, and three saltwater invertebrates. Eight records were found in HSDB.

The most sensitive chronic effects were seen in a freshwater fish, the rainbow/Donaldson trout (*Oncorhynchus mykiss*). The 552-hour study (23 days) and a 648-hour study (27 days) each reported an LC_{50} for mortality of 70 ppb. A 2040-hour lowest observed effect concentration (LOEC) (85 days) for growth effects of 99.4 ppb was also reported for rainbow/Donaldson trout.

Freshwater invertebrates are less sensitive than freshwater fish. The most sensitive freshwater invertebrate was the water flea (*Daphnia magna*). The 504-hour study (21 days) reported a NOEC for mortality of 1050 ppb. A 168-hour LC_{50} (7 days) for mortality of 2,608 ppb was also reported for water flea.

8.6 THM: Bromodichloromethane

8.6.1 Mammalian Toxicity

Data on the toxicity or carcinogenicity of bromodichloromethane in humans are inadequate. No epidemiologic studies examining bromodichloromethane alone were located. NTP (1986) has conducted both subchronic and chronic studies on the toxicity and carcinogenicity of bromodichloromethane. Based on these studies, U.S. EPA has derived an RfD of $2x10^{-2}$ mg/kg-day based on renal cytomegaly in male mice and an oral slope factor of $6.2x10^{-2}$ per (mg/kg-day) and drinking water unit risk of $1.8x10^{-6}$:g/Liter for bromodichloromethane (U.S. EPA, 2004b). In 1999, IARC classified bromodichloromethane as possibly carcinogenic to humans, based on no data in humans and sufficient evidence in animals (IARC, 2001). ACGIH (2000) has not evaluated the toxicity or carcinogenicity of bromodichloromethane.

8.6.1.1 Noncancer Effects

Data on the toxicity of bromodichloromethane in humans are lacking. No epidemiologic studies examining bromodichloromethane alone were located. NTP (1986) conducted a lifetime carcinogenicity bioassay of bromodichloromethane in rats and mice. Groups of 50 male and 50 female rats or 50 male and 50 female mice were exposed to bromodichloromethane in corn oil once per day, 5 days/week for 102 weeks. Rats of either sex were exposed to 0, 50, or 100 mg/kg per exposure, while male mice were exposed daily to 0, 25, or 50 mg/kg and groups of 50 female mice were exposed daily to 0, 75, or 150 mg/kg. Final survival of dosed female mice was reduced compared with controls. Final mean body weights of dosed female mice and high-dose male and female rats were 75-91% that of vehicle controls. Compound-related nonneoplastic lesions included cytomegaly and tubular cell hyperplasia of the kidney and fatty metamorphosis of the liver in male rats; eosinophilic cytoplasmic change, clear cell change, focal cellular change, and fatty metamorphosis of the liver and tubular cell hyperplasia of the kidney in female rats; fatty metamorphosis of the liver, renal cytomegaly, and follicular cell hyperplasia of the thyroid gland in male mice; and follicular cell hyperplasia of the thyroid gland in female mice. A LOAEL of 17.9 mg/kg-day (25 mg/kg-day x 5 days/7 days) based on renal cytomegaly in male mice, an effect considered minimal in the absence of data demonstrating renal functional impairment, is indicated by these results.

In a subchronic bioassay conducted by NTP (1986), male and female rats received doses of 19 to 300 mg/kg-day, male mice received doses of 6.25 to 100 mg/kg-day, and female mice received doses of 25 to 400 mg/kg-day for 5 days/week. Five of 10 male rats and 2/10 female rats at 300 mg/kg-day died. Final body weights of male and female rats in the 150 or 300 mg/kg-day treatment groups, male mice receiving 100 mg/kg-day and female mice receiving 200 or 400 mg/kg-day, were lower than those of controls. Degeneration of the liver and degeneration and necrosis of the kidney were seen in high-dose male rats; liver lesions were observed in high-dose female rats and in female mice at 200 or 400 mg/kg-day, and kidney lesions were seen in male mice at 100 mg/kg-day. These data define a NOAEL of 35.7 mg/kg-day (50 mg/kg-day x 5 days/7 days), a dose above which produced kidney lesions and depressed body weight in male mice. Because the chronic study used more animals/dose, was of longer duration, and presented more complete data, more confidence is placed in the chronic LOAEL than in the subchronic NOAEL.

8.6.1.2 Carcinogenic Effects

Data on the carcinogenic effects of bromodichloromethane in humans are lacking. No epidemiologic studies examining bromodichloromethane alone were located. A two-year carcinogenicity study of bromodichloromethane by NTP (1986) found sufficient evidence for carcinogenic effects. Groups of 50 male and 50 female rats or 50 male and 50 female mice were exposed to bromodichloromethane in corn oil once per day, 5 days/week for 102 weeks. Rats of either sex were exposed to 0, 50, or 100 mg/kg per exposure, while male mice were exposed daily to 0, 25, or 50 mg/kg, and groups of 50 female mice were exposed daily to 0, 75, or 150 mg/kg. Survival was reduced 52%, 26%, and 30% in the control, low-dose, and high-dose females, respectively, after week 84; the mortality was associated with ovarian abscesses.

In male and female rats, the incidence of certain kidney cancers was statistically significantly increased only in the high-dose (100 mg/kg-day) groups. In female mice, an increase in hepatic tumors was also noted, but the significance of this finding to humans is questionable. In male mice, the incidence of certain cancers of the kidney was significantly increased in the high-dose (50 mg/kg-day) group (1/46, 2/49 and 9/50 in the control, low-dose and high-dose groups, respectively). Based on the tumor incidence in male mice, EPA calculated an oral slope factor of 6.2×10^{-2} per (mg/kg-day) and drinking water unit risk of 1.8×10^{-6} :g/Liter for bromodichloromethane. Based on inadequate data in humans and sufficient evidence in two animal species, bromodichlormethane is classified as category B2, a probable human carcinogen (U.S. EPA, 2004b). Four other carcinogenicity studies support the carcinogenicity assessment. Bromodichloromethane was also reported to be mutagenic in certain bacterial strains (U.S. EPA, 2004b).

8.6.2 Aquatic Ecotoxicity

There is no ecotoxicity data available for bromodichloromethane.

8.7 THM: Dibromochloromethane

8.7.1 Mammalian Toxicity

Data on the toxicity or carcinogenicity of dibromochloromethane in humans are inadequate. No epidemiologic studies examining dibromochloromethane alone were located. NTP (1985) has conducted subchronic and chronic studies of dibromochloromethane in rats and mice. From these studies, U.S. EPA has derived an RfD of 2×10^{-2} mg/kg-day, an oral slope factor of 8.4×10^{-2} per (mg/kg-day), and a drinking water unit risk of 2.4×10^{-6} per (:g/L) (U.S. EPA, 2004). Dibromochloromethane is classified as category C (possible human carcinogen) by U.S. EPA, and has not been evaluated by IARC (2001) or ACGIH (2000).

8.7.1.1 Noncancer Effects

Rats and mice (10/sex/dose) were administered dibromochloromethane at 0, 15, 30, 60, 125, or 250 mg/kg-day by gavage for five days/week for 13 weeks. At 250 mg/kg-day, male mice showed increased incidence of toxicity of the liver and kidneys. Both sexes of rats showed increased incidence of liver and kidney toxicity, as well salivary gland inflammation at 250 mg/kg-day. Liver toxicity also occurred in lower-dose male rats. Statistical analysis indicated that 30 mg/kg-day is the no observed effect level (NOEL) for liver toxicity (NTP, 1985). This effect was used as the basis for the RfD (U.S. EPA, 2004b).

In the chronic bioassay portion of the NTP (1985) study, 50 rats of each sex were administered dibromochloromethane at 0, 40, or 80 mg kg/day by gavage for 104 weeks and 50 B6C3F1 mice of each sex were similarly treated with 0, 50, or 100 mg DBCM/kg/day for 105 weeks. Treatment was five days/week. A dose-related increase in histologic changes in the liver of treated rats of both sexes was reported. Treated female rats also had higher incidences of kidney injury. Treated mice of both sexes exhibited higher incidences of liver damage. The incidence of kidney injury was increased in dosed males, and thyroid follicular cell hyperplasia was increased in dosed female mice. The LOAELs for this portion of the study are 40 mg/kg-day for rats and 50 mg/kg-day for mice. The results from these chronic bioassays support the subchronic studies used as a basis of the RfD. In this case, the choice of the subchronic NOAEL as the basis of the RfD rather than the chronic LOAEL—either choice normally requires a 1000 uncertainty factor (UF), and the resulting RfDs are similar—reflects the slightly greater confidence in the subchronic NOAEL versus the chronic LOAEL that was associated with several adverse effects.

8.7.1.2 Carcinogenic Effects

In a two-year carcinogenicity study, 50 rats/sex/dose were treated by gavage with dibromochloromethane (>98% pure) in corn oil at 0, 40, or 80 mg/kg, 5 days/week for 104 weeks (NTP, 1985). Groups of mice (50/sex/dose) were similarly gavaged at doses of 0, 50, or 100 mg/kg, 5 days/week for 105 weeks. In rats, the final survival rates of all groups were comparable (approximately 76%); mean body weights were also comparable between dose groups in each sex, except for a decrease in the high-dose males after week 20. No compound-related clinical signs were seen, and no evidence for carcinogenicity was seen in rats under these study conditions.

In mice, the mean body weights of both male and female high-dose groups were lower than those of their respective controls throughout the study. Mean body weight of the low-dose male mice was lower than the control group after both low-dose groups received an overdose of chemical in week 58. Survival data from the study were impacted by an accidental overdose during the treatment period. In female mice, the incidence of two forms of liver cancer was statistically significantly increased in the high-dose group. In high-dose male mice, there was a similar increase in the incidence of liver cancer, but only a specific form. Under the conditions of this study, NTP (1985) determined that there was equivocal evidence of carcinogenicity of dibromochloromethane in male mice, and some evidence of carcinogenicity in female mice. Other chronic rodent studies have not reported liver cancer following exposure to dibromochloromethane (U.S. EPA, 2004b).

In support of the cancer data, dibromochloromethane has been shown to produce mutations in certain bacterial strains and in bone marrow cells of mice treated orally (U.S. EPA, 2004b).

8.7.2 Aquatic Ecotoxicity

A total of 15 unique records were found in AQUIRE. Fourteen records were excluded from the analysis due to insufficient or incompatible information, e.g., no study endpoint. The remaining record was grouped as a freshwater fish. The database contained information on one species of freshwater fish and one freshwater invertebrate. Additional aquatic toxicity studies were summarized in HEED/HEEP.

The most sensitive acute freshwater effects were seen in common, mirror, colored, carp (*Cyprinus carpio*). The 0.75-hour study reported an LC_{50} for mortality of 34,000 ppb. None of the records contained information on freshwater invertebrates. None of the records contained information on saltwater fish or invertebrates. None of the records contained information on chronic saltwater or freshwater endpoints.

8.8 THM: Bromoform

8.8.1 Mammalian Toxicity

Data on the toxicity and carcinogenicity of bromoform in humans are inadequate. NTP (1988, 1989) has conducted subchronic (toxicity) and chronic (carcinogenicity) studies on bromoform by the oral route. EPA (U.S. EPA, 2004b) derived an RfD of $2x10^{-2}$ mg/kg-day for hepatic lesions in the subchronic NTP bioassay (NTP, 1989). Based on the chronic NTP study (NTP, 1988), EPA derived an oral slope factor of $7.9x10^{-3}$ per (mg/kg-day) and a drinking water unit risk of $2.3x10^{-7}$ per (:g/L) for bromoform (U.S. EPA, 2004b). IARC reported that the carcinogenicity of bromoform was not classifiable, based in inadequate data in humans and limited evidence of carcinogenicity in animals. ACGIH (2000) has classified bromoform as category A3: confirmed animal carcinogen with unknown relevance to humans.

8.8.1.1 Noncancer Effects

NTP (1989) conducted a subchronic bioassay of bromoform toxicity. In a subchronic study, rats (10/sex/group) and mice (10/sex/group) received oral doses of bromoform at 0, 12, 25, 50, 100, or 200 mg/kg or 0, 25, 50, 100, 200, or 400 mg/kg, respectively. Complete histology was conducted on high-dose and vehicle control groups of both species. Females of both species did not show any dose-related effects. The male mice showed fatty metamorphosis of the liver at doses of 200 and 400 mg/kg. The only effect reported for male rats was a dose-related increase in pathology of the liver, which was statistically elevated relative to controls. Therefore, 25 mg/kg is the NOEL for rats (NTP, 1989).

8.8.1.2 Carcinogenic Effects

In a gavage study (NTP, 1988), rats (50/sex/group) and mice (50/sex/group) were administered bromoform in corn oil by gavage five days/week for two years at 0, 100, or 200 mg/kg (rats and female mice) or 0, 50, or 100 mg/kg (male mice). Decreased body weight and survival in rats and female mice suggest that the maximum tolerated dose was reached. In male rats, mean body weight was decreased in the high- and low-dose groups by 12–28% and 5%–14%, respectively. Survival was significantly lower in the high-dose males after week 91. In female rats, body weight was decreased in the high-dose group by 10–25%. In male mice, body weight and survival were comparable to controls. In female mice, however, body weight was decreased in the high- and low-dose groups by 5–16% and 6–11%, respectively; survival was significantly lower in both dose groups after week 77. Cancerous lesions were observed in the large intestine (colon or rectum) of male rats (0/50, 0/50, 3/50) and female rats (0/50, 1/50, 8/50) rats. No treatment-related tumors were observed in mice at either dose level. Under the conditions of this study, the NTP judged that there was clear evidence of carcinogenicity for female rats, some evidence of carcinogenicity for male rats, and no evidence of carcinogenicity for male and female mice. Other research has reported that bromoform may possess tumor promotion activity in the skin and liver (Pereira, 1983). Supporting these data, a number of studies suggest that bromoform is genotoxic, producing mutations in bacterial strains and chromosomal damage in animals.

8.8.2 Aquatic Ecotoxicity

A total of 44 unique records were found in AQUIRE. There were 26 records excluded from the analysis due to insufficient or incompatible information, e.g., no study endpoint. The remaining 17 records were grouped as follows: four as freshwater fish, three as freshwater invertebrates, seven as saltwater fish, and three as saltwater invertebrates. The database contained information on two species of freshwater fish, one freshwater invertebrate, two as saltwater fish, and two as saltwater invertebrates. Four records were found in HSDB. Additional aquatic toxicity studies were summarized in HEED/HEEP.

The most sensitive acute freshwater effects were seen in a fish, the bluegill (*Lepomis macrochirus*). The 96-hour study reported an LC_{50} for mortality of 29,000 ppb. Freshwater invertebrates are less sensitive than freshwater fish. The most sensitive freshwater invertebrate was the water flea (*Daphnia pulex*). The 96-hour study reported an LC_{50} for mortality of 44,000 ppb. None of the records contained information on chronic freshwater endpoints.

The most sensitive acute saltwater effects were seen in an invertebrate, the eastern oyster larvae (*Crassostrea virginica*). The 48-hour study reported an LC_{50} of 1,000 ppb. Saltwater fish are less sensitive than saltwater invertebrates. The most sensitive saltwater fish was the sheepshead minnow (*Cyprinodon variegatus*). The 96-hour study reported an LC_{50} of 7,100 ppb. The most sensitive chronic saltwater effects were also seen in the sheepshead minnow (*C. variegatus*). The maximum allowable/acceptable toxicant concentration (MATC) was reported as >4,800 to <8,500 ppb.

8.9 HAA: Monochloroacetic Acid

8.9.1 Mammalian Toxicity

Data on the toxicity and carcinogenicity of monochloracetic acid in humans are inadequate.

8.9.1.1 Noncancer Effects

Acute (short-term) inhalation or dermal exposure to chloroacetic acid may cause severe damage to the skin and mucous membranes in humans. Chloroacetic acid irritates and may burn the skin, eyes, and respiratory tract. Death may occur if more than 3% of the skin is exposed to this chemical (DOH 1993; Merck 1989; U.S. EPA 1985). Depression of the central nervous system may occur in humans following acute inhalation exposure (U.S. EPA 1985).

Mice acutely exposed by ingestion have exhibited neurological dysfunction (U.S. EPA 1998). Acute exposure by ingestion of chloroacetic acid may interfere with essential enzyme systems in the body and cause intestinal perforation and peritonitis in humans (DOH 1993; Merck 1989; U.S. EPA 1985).

Tests involving acute exposure of animals, such as the LC_{50} and LD_{50} tests in rats, mice, and guinea pigs, have demonstrated chloroacetic acid to have extreme toxicity from inhalation and moderate to high acute toxicity from ingestion (DOH 1993).

U.S. EPA's Office of Air Quality Planning and Standards, for a hazard ranking under Section 112(g) of the Clean Air Act Amendments, considers chloroacetic acid to be a "high concern" pollutant based on severe acute toxicity (U.S. EPA 1994b).

No information is available on the chronic (long-term) effects of chloroacetic acid in humans. Damage to the respiratory tract, including inflammatory changes in the respiratory organs, inflammatory lesions of the nasal mucosa, metaplasia of the olfactory epithelium, and respiratory congestion, have been observed in rodents chronically exposed to chloroacetic acid by inhalation, orally, and via gavage (U.S. EPA 1998; NTP 1992). Chronic exposure via gavage has resulted in myocarditis (inflammation of the muscular tissue of the heart wall) and mortality due to myocardial failure in rats and hepatic vacuolar degeneration in mice (U.S. EPA 1998; NTP 1992).

EPA has not established an RfC for chloroacetic acid. U.S, EPA has calculated a provisional RfD of 0.002 mg/kg/d for chloroacetic acid (NTP 1992).

8.9.1.2 Carcinogenic Effects

No information is available on the carcinogenic effects of chloroacetic acid in humans (U.S. EPA 1998). Chloroacetic acid was not found to be tumorigenic to mice when administered via gavage or by subcutaneous injection or when applied to the skin (U.S. EPA 1998). In an NTP study, no statistically significant increases in tumor incidences were reported in rats and mice exposed to chloroacetic acid via gavage (NTP 1992).

U.S. EPA has classified chloroacetic acid as a Group D, not classifiable as to human carcinogenicity (U.S. EPA 1998).

8.9.2 Aquatic Ecotoxicity

There was one study for monochloroacetic acid found in the AQUIRE database that reported on both acute and chronic effects on the water flea (Kuhn et al., 1989). The 48-hour study reported an EC50 of 77,000 µg/L based on an immobility endpoint. The 21-D NOEC study reported a value of 32,000 µg/L based on a reproduction endpoint.

8.10 HAA: Dichloroacetic Acid

8.10.1 Mammalian Toxicity

Data on the toxicity and carcinogenicity of dichloroacetic acid in humans is inadequate. EPA (U.S. EPA, 2004b) derived an RfD of 4×10^{-3} mg/kg-day based on lesions in the testes, cerebrum, cerebellum, and liver in an oral, subchronic study in beagle dogs (Cicmanec et al., 1991). Based on the study of DeAngelo et al. (1999), an oral slope factor of 5×10^{-2} per (mg/kg-day) and a drinking water unit risk of 1.4 x 10^{-6} per (μg/L) were derived (U.S. EPA, 2004b). The primary source of exposure to dichloroacetic acid in humans is as a by-product of chlorination of drinking water. Since the early 1970s, numerous epidemiological studies have attempted to assess the relationship between the chlorination by-products in drinking water and several different human cancers. None of these studies has shown an association of exposure to dichloroacetic acid with an increased incidence of site-specific cancer. Agency for Toxic Substances and Disease Registry (ATSDR) has not published a toxicological profile on dichloroacetic acid (ATSDR, 2001). ACGIH has not evaluated dichloroacetic acid (2000). According to IARC, the available data were too limited to form the basis for an evaluation of the carcinogenicity of dichloroacetic acid to humans. Thus dichloroacetic acid is not classifiable as to its carcinogenicity to humans (Group 3) (IARC, 2001).

8.10.1.1 Noncancer Effects

Juvenile beagle dogs (5/sex/dose) were administered dichloroacetic acid at 12.5 to 72 mg/kg-day in capsules for 90 days (Cicmanec et al., 1991). At the high dosage, dyspnea, partial paralysis of the hind limbs, and ocular effects were observed. A decrease in food and water intake was observed. High-dose males showed a decrease in body weight (16%). High-dose females exhibited a decrease in body weight gain (9%). Three deaths in the high-dose group were attributed to pneumonia and dehydration. Hematological effects were also observed in the high-dose group.

Relative liver weights significantly increased in all dose groups, and absolute liver weight increased in all but high-dose males. Relative kidney weights increased in mid- and high-dose males and females, and relative lung weights increased in high-dose males and females. Mild to moderate kidney lesions, mild vacuolar changes, inflammation, and hemosiderosis in the liver, and chronic inflammation in the pancreas were observed in treated animals. Mild vacuolization of white myelinated tracts in the cerebrum and cerebellum in animals of all dose groups as well as mild vacuolar changes in the medulla and spinal cord of some mid- and high-dose males were observed. Microscopic testicular lesions were reported in males at all dose levels with severity increasing with dose. The testicular changes were apparent in 4/5 males at the low dose and in all animals in the mid- and high-dose groups (Cicmanec et al., 1991).

Testicular degeneration, mild to moderate hepatic vacuolization in both sexes, and mild vacuolization of the myelinated white tracts of the cerebrum and cerebellum in males were observed in the low-dose group. A LOAEL of 12.5 mg/kg-day was determined, based on visual organ effects (neurological changes, hepatic vacuolization, and testicular effects) and increases in liver weight (Cicmanec et al., 1991). This LOAEL was used as the basis for the RfD (U.S. EPA, 2004b).

Dichloroacetic acid is metabolized in humans and experimental animals, and oxalate, thiodiacetic acid, and unchanged dichloroacetic acid are excreted in urine. Clearance is decreased in humans after repeated administration. Species differences in the clearance of dichloroacetic acid are observed in rodents: clearance in rats is much slower than in mice. Dichloroacetic acid induces peroxisome proliferation in the livers of both rats and mice (Larson and Bull, 1992). No data were available on the effects of dichloroacetic acid on human reproduction. In rats and dogs, testicular degeneration can occur after exposure to this compound. The development of the hearts, major vessels, and kidneys of rats can be affected by exposure *in utero* (Stacpole, 1990).

Moore et al. (1979) administered daily oral doses of dichloroacetic acid at 50 mg/kg to two male patients (ages 8 and 21) with severe familial hypercholesterolemia for 16 weeks. Serum cholesterol levels dropped but did not reach normal levels. No adverse clinical or laboratory signs related to dichloroacetic acid treatment were detected in the eight-year-old individual.

8.10.1.2 Carcinogenic Effects

In seven separate studies in mice, (DeAngelo et al., 1999; DeAngelo et al., 1991; Daniel et al., 1992; Herren-Freund et al., 1987; Bull et al., 1990; Pereira, 1996; Pereira and Phelps, 1996), and three studies in rats (DeAngelo et al., 1996; Richmond et al., 1995), oral administration of dichloroacetic acid increased the incidence of liver cancer.

DeAngelo et al. (1999) exposed male $B6C3F_1$ mice to 0, 0.05, 0.5, 1, 2, or 3.5 g/L of dichloroacetic acid (DCA) in drinking water for 90–100 weeks. This corresponded to mean daily doses of 0, 8, 84, 168, 315, or 429 mg/kg-day, respectively. The cumulative incidence of hepatocellular carcinomas significantly increased in animals exposed to 168 mg/kg-day (71%), 315 mg/kg-day (95%), and 429 mg/kg-day (100%) when compared to controls (26%). Hepatocellular carcinoma multiplicity (tumor/animal) significantly increased in all treatment groups compared to the control group. The cumulative incidence of hepatocellular adenomas was also significantly increased in animals exposed to 168 mg/kg-day and up as compared to controls. Hepatocellular adenoma multiplicity (tumor/animal) significantly increased with doses of 168 mg/kg-day and higher as compared to controls. Hepatic peroxisome proliferation increased in the high-dose group but did not correlate with liver tumor response. The severity of hepatotoxicity increased with increasing dose. Below 168 mg/kg-day, hepatotoxicity was mild. Thus, neither peroxisome induction nor chemically-sustained cell proliferation were necessary for the development of tumors. Based on the study of DeAngelo et al. (1999), an oral slope factor of $5x10^{-2}$ per (mg/kg-day) and a drinking water unit risk of $1.4x10^{-6}$ per (µg/L) were derived (U.S. EPA, 2004b) for dichloroacetic acid. Based on inadequate data in humans and sufficient evidence in two animal species, dichloroacetic acid is classified as category B2; a probable human carcinogen (U.S. EPA, 2004b).

Increases were observed in hepatocellular adenomas and carcinomas in male mice receiving dichloroacetic acid (DeAngelo et al., 1991). Groups of 50 male mice received drinking water containing neutralized dichloroacetic acid at levels of 0.05, 0.5, and 5 g/L (corresponding to time-weighted mean intake of 7.6, 77, and 486 (mg/kg)/day. Statistically significant increases in the relative liver weights were seen in the 0.5-, 3.5-, and 5-g/L groups (36%, 132%, and 250% greater than control) at 60 weeks. Hepatocellular adenomas in the 0.5- and 3.5-g/L groups were first seen (in sacrificed animals) at 45 and 30 weeks, respectively, while hepatocellular carcinomas in the 0.05-g/L group were first seen at 60 weeks. In the high-dose group, hepatocellular adenomas and carcinomas were first seen at 45 weeks. At the 60-week sacrifice, hepatocellular adenomas were present in 12/12 (100%) and 24/30 (80%) mice receiving 3.5 and 5 g/L, respectively, and carcinomas were found in 8/12 (67%) and 25/30 (83%) mice, respectively. The incidences of combined hepatocellular tumors showed a significant statistical increase, 100% and 90% for the 3.5- and 5-g/L groups, respectively, and the multiplicity of liver tumors was 4 and 4.5/mouse, respectively. Hyperplastic liver nodules were noted in 7/12 (58%) and 25/30 (83%) of the 3.5- and 5-g/L groups, respectively (incidence not significantly elevated).

Male mice (33/group) were treated with 0 or 0.5 g/L dichloroacetic acid 88 (mg/kg)/day in drinking water for 104 weeks (Daniel et al., 1992). Nonneoplastic changes in the liver of the 0.5-g/L animals included hepatocellular necrosis (33% incidence), hyperplasia (33%), and cytomegaly (92%), compared with controls (less than or equal to 5%). Hepatocellular carcinomas developed in 15/24 (63%) treated mice compared with 2/20 (10%) controls, and hepatocellular adenomas occurred in 10/24 (42%) treated mice compared with 1/20 (5%) controls. The incidence of combined hepatocellular tumors was 18/24 (75%) in 0.5-g/L mice compared with 3/20 (15%) in controls. Individual type and combined tumor incidences showed a significant statistical elevation in the treated group compared with controls. Hyperplastic liver nodules were observed in 2/24 (8%) of the treated males but none in the controls (elevation not statistically significant).

Herren-Freund et al. (1987) investigated the initiation/promotion potential of dichloroacetic acid in male mice (study groups of 26 and 32). Mice were either given 5 g/L dichloroacetic acid or pretreated with a single intraperitoneal dose of 2.5 mg/kg ethylnitrosourea (ENU) at 15 days of age and then given 2 or 5 g/L dichloroacetic acid in the drinking water (400 or 1000 mg/kg-day as calculated by the authors) during the period from 4–65 weeks of age. The negative control group (66 animals) received 2 g/L NaCl in drinking water. Animals were sacrificed after 61 weeks of exposure. A significant decrease in the final mean body weight was observed in the 5-g/L groups (uninitiated and initiated) (both 92% of control), while a significant increase in the relative liver weight was found in all treatment groups (140–190% of control). In the uninitiated group receiving 5 g/L dichloroacetic acid, the incidences were 25/26 (96%) for hepatocellular adenomas and 21/26 (81%) for hepatocellular carcinomas compared with 2/22 and 0/22, respectively, for the controls. For the initiated groups, the incidence was 22/29 (76%) and 31/32 (97%) for hepatocellular adenoma in the 2- and 5-g dichloroacetic acid per liter groups, respectively, compared with 2/22 (9%) in the controls. The incidence was 19/29 (66%) and 25/32 (72%) for hepatocellular carcinomas in the initiated 2- and 5-g dichloroacetic acid per liter groups, respectively, compared with 1/22 (4.5%) in the controls. All tumor incidences in the treated groups showed a significant elevation. Hyperplastic nodules were not reported in the treated animals, and no other tumor sites were found.

Bull et al. (1990) also demonstrated that dichloroacetic acid was carcinogenic in male mice. In the study, males received 1 g/L dichloroacetic acid [170 (mg/kg)/day] in drinking water for 52 weeks (11 animals) or 2 g/L (340 mg/kg-day) (11–24 animals) for 37 or 52 weeks. Ten females received 2 g/L dichloroacetic acid for 52 weeks. The control group consisted of 61 males (sacrificed at 15, 24, 37, or 52 weeks) and 10 females (sacrificed at 52 weeks). No effects of treatment on survival or body weight were observed. Food and water consumption data were not reported. A significant increase in the relative liver weight was seen in 1 g/L males (41% increase from control), 2 g/L males (128% increase), and 2 g/L females (88% increase) compared with controls at 52 weeks. Nonneoplastic lesions included degenerative changes (multifocal areas of necrosis with frequent lymphocyte infiltration), cytomegaly and glycogen accumulation in the liver of all treated animals (incidence data were not provided). The high-dose mice also exhibited basophilic foci of cellular alteration in the livers at the 24- and 37-week sacrifice. Hepatocellular adenomas and carcinomas both developed only in the 2 g/L males, in 2/11 and 2/24 males treated for 37 and 52 weeks, respectively. Hepatocellular carcinomas alone developed in 5/24 males treated for 52 weeks. Tumor incidences in all treated groups were statistically significantly elevated. Hyperplastic nodules were observed in 6/11 and 9/24 males treated with 2 g/L dichloroacetic acid for 37 and 52 weeks, respectively, and in 3/10 females treated with 2 g/L for 52 weeks. The appearance of hyperplastic nodules in the treated females suggests that tumors might develop in female mice if dichloroacetic acid exposure were continued beyond 52 weeks. Tumors were not found in any other organs. Since the total exposure duration in this study did not exceed 52 weeks, this study may not have evaluated animals for an adequate length of time to observe the full carcinogenic potential of dichloroacetic acid in mice.

Female B6C3F$_1$ mice were administered 40, 115, or 330 mg/kg-day DCA in drinking water from seven to eight weeks of age to sacrifice at 360 or 576 days (~ 51 or 82 weeks) of exposure (Pereira, 1996). Significant increases in the percentage of animals with altered hepatocyte foci and liver adenomas were seen in the 115 and 330 mg/kg-day dose groups as follows: after 51 weeks, 40.0% with foci and 35% with adenomas at 330 mg/kg-day; after 82 weeks, 39.3% with foci and 25% with adenomas at 115 mg/kg-day, and 89.5% with foci and 84.2% with adenomas at 330 mg/kg-day. A significant increase in the percentage of animals with liver carcinomas was seen only in the 330 mg/kg-day group after 82 weeks of exposure.

Pereira and Phelps, (1996) initiated liver tumors in female B6C3F$_1$ mice with 25 mg/kg methylnitrosourea (MNU); the mice were then administered 50, 167, or 468 mg/kg-day DCA in their drinking water from age seven weeks to sacrifice at 31 or 52 weeks later to determine tumor promotion by DCA. Significant increases in the percentage of animals with liver adenomas were seen in the 468 mg/kg-day group after 31 and 52 weeks of exposure. A significant increase in the percentage of animals with altered hepatocyte foci was also seen after 31 and 52 weeks of exposure. When exposure to 468 mg/kg-day DCA was terminated after 31 weeks, followed by a 21-week recovery period, the yield of foci of altered hepatocytes and tumors decreased, indicating that continued existence of these lesions was dependent on continuous exposure to DCA. The tumor-promoting activity of DCA exhibited a second-order relationship to drinking water concentration, so that a sharp rise in potency was seen between 167 and 468 mg/kg-day.

The carcinogenic potential of DCA has also been evaluated in the rat model. In one study, 28-day-old male F344 rats were given drinking water containing DCA at concentrations of 0 (78/group), 0.05 (60/group), 0.5 (60/group) or 5.0 g/L (78/group) (DeAngelo et al., 1996). Another group was provided water containing 2.0 g/L NaCl (50/group). Body weights and water consumption were measured throughout the study. All animals were treated for 100 weeks, except for the 5.0 g/L group, which began to exhibit signs of peripheral neuropathy. When the neuropathy did not reverse or diminish after decreasing the dose, the animals were sacrificed at 60 weeks and excluded from the report. Based on measured water intake in the 0, 0.05 and 0.5 g/L groups, the time-weighted average doses were 0, 3.6, and 40.2 mg/kg-day respectively. Interim sacrifices for each dose group were performed at 15, 30, 45, 60, and 100 weeks; the NaCl control group was sacrificed at 104 weeks. No differences were observed in water consumption, final body weight, absolute or relative liver, kidney, or spleen weight, at doses of 3.6 or 40.2 mg/kg-day at any time point. Absolute and relative testicular weights were significantly increased at the 40.2 mg/kg-day dose at final sacrifice. Increased hepatocellular vacuolization was detected, but there was no increase in hepatocyte proliferation at any dose group. At a dose of 40.2 mg/kg-day DCA, there was a statistically significant increase in the cumulative incidence of combined hepatocellular neoplasia and total proliferative lesions in the liver compared to controls. This was not observed at the lower dose of 3.6 mg/kg-day. Tumor multiplicity was significantly increased in the 40.2 mg/kg-day group as compared to controls. There was also a significant increase in combined hepatocellular neoplasia and total proliferative lesions.

In another study, male F344 rats were exposed to DCA concentrations of 2.5 g/L DCA in their drinking water (78/group) or to deionized water (78/group) (DeAngelo et al., 1996). The concentration of DCA was lowered to 2 g/L at five weeks, to 1.5 g/L at eight weeks, and to 1.0 g/L at 26 weeks. This corresponded to a time-weighted average concentration of 1.6 g/L and a time-weighted average dose of 139 mg/kg-day over the 103-week exposure period. Interim sacrifices for each dose group were performed at 14, 26, 52, 78, and 103 weeks. There was a 73% decrease in mean final body weight of DCA-exposed animals compared to the control group. Absolute testes weight decreased. Hepatic tumor incidence significantly increased in exposed animals compared to controls. Tumor multiplicity was also significantly increased in this group compared to controls. Other tumors examined were not increased over control values.

Male F344 rats were administered time-weighted average concentrations of 0, 0.05, 0.5, or 2.4 g/L (0, 4, 40, or 296 mg/kg-day) DCA in drinking water, followed by sacrifice at intervals for up to 104 weeks by Richmond et al. (1995). No hepatoproliferative lesions were seen in the 4 mg/kg-day group, and the negative control group had only 4% hepatic adenomas. The 40 mg/kg-day group had 10% hyperplastic nodules, 21% hepatic adenomas, and 10% hepatocarcinomas after 104 weeks, while the 296 mg/kg-day group had 70% hyperplastic nodules, 26% hepatic adenomas, and 4% hepatocarcinomas after terminal sacrifice at 60 weeks. Increased numbers of altered hepatocyte foci were also seen in the 4 and 40 mg/kg-day groups, but the differences were significant only in animals from the 45-week sacrifice.

U.S. EPA (1991) reported on the carcinogenicity of dichloroacetic acid in female mice. Following exposure to 0, 0.5 or 3.5 g/L dichloroacetic acid in drinking water for 104 weeks, the high-dose group had a 100% hepatocellular tumor incidence and a tumor multiplicity of 8.36 tumors/animal. Females receiving 0.5 g/L dichloroacetic acid had a tumor incidence of 20% and a tumor multiplicity of 0.2 tumors/animal. These values were compared with an incidence of 7.7% and a multiplicity of 0.1 tumors/animal in the control group. It was concluded that, although statistical analysis was not performed for the study, it appeared that female mice might be less sensitive than male mice.

The evidence for induction of DNA strand breaks in cells exposed to dichloroacetic acid *in vivo* and *in vitro* was inconclusive or negative (Nelson and Bull, 1988; DeAngelo et al., 1989; Chang et al., 1992). The results of assays for mutagenesis in bacteria were inconsistent (NTP, 1991a, 1991b; Ono et al., 1991). According to IARC (2001), the majority of data on mutagenicity and genotoxicity in a number of *in vitro* and *in vivo* test systems indicates that dichloroacetic acid is not mutagenic. No data were found on the genotoxic effects of dichloroacetic acid in eukaryotic organisms (IARC, 2001).

8.10.2 Aquatic Ecotoxicity

A total of 11 unique records were found in AQUIRE. Ten records were excluded from the analysis due to insufficient or incompatible information, i.e., no study endpoint. The remaining record was grouped as a saltwater invertebrate.

The most sensitive acute saltwater effects were seen in an invertebrate, the harpacticoid copepod (*Daphnia magna*). The 96-hour study reported an LC_{50} for mortality of 23,000 ppb. None of the records contained information on saltwater fish or chronic endpoints. None of the records contained information on freshwater fish or invertebrates.

8.11 HAA: Trichloroacetic Acid

8.11.1 Mammalian Toxicity

8.11.1.1 Noncancer Effects

Trichloroacetic acid is a skin and eye irritant and also affects the kidneys and liver of experimental animals. It was shown to be a liver carcinogen in a single study in mice (ACGIH 1991). When trichloroacetic acid was instilled into the eye of rabbits, severe irritation was observed characterized by severe and extensive loss of epithelium and endothelium, infiltration and hemorrhage about the limbal vessels, infiltration and hyperemia of the iris, and damage to the anterior layers of the lens (ACGIH, 1991). Trichloroacetic acid is a mild skin irritant in rabbits (NIOSH, 1991). The oral LD_{50} in rats is 400 mg/kg (NIOSH 1991). Animals poisoned by oral doses of trichloroacetic acid quickly passed into a state of narcosis or semi-narcosis, and within 36 hours they either recovered completely or died (ACGIH, 1991). Male rats administered 50, 500, or 5,000 ppm trichloroacetic acid in their drinking water for 90 days developed a significant increase in relative kidney weight, a significant increase in hepatic peroximal beta oxidation activity, and focal areas of hepatocellular hypertrophy (ACGIH, 1991). Trichloroacetic acid ingestion had no effect on serum biochemical parameters or on parameters of immune function (ACGIH, 1991). Trichloroacetic acid was not mutagenic in the Ames assay in the presence and absence of metabolic activation. It showed some activity in *in vivo* tests of genotoxicity (ACGIH, 1991).

8.11.1.2 Carcinogenic Effects

The available data were too limited to form the basis for an evaluation of the carcinogenicity of trichloroacetic acid to humans. In a chronic study, mice were administered trichloroacetic acid in the drinking water at concentrations of 2,000 or 5,000 ppm for 61 weeks. Both dose levels of trichloroacetic acid induced hepatocellular carcinomas and adenomas (NLM, 1992; ACGIH, 1991). Trichloroacetic acid was not carcinogenic in rats (details of this study were not provided) (ACGIH, 1991).

8.11.2 Aquatic Ecotoxicity

A total of 75 unique records were found in AQUIRE. Forty-three records were excluded from the analysis due to insufficient or incompatible information, e.g., no study endpoint. The remaining 23 records were grouped as follows: 23 as freshwater fish, seven as freshwater invertebrates, one as saltwater fish, and one as saltwater invertebrates. The database contained information on nine species of freshwater fish, four freshwater invertebrate, 12 saltwater fish, and one saltwater invertebrates.

The most sensitive freshwater fish species was from the Salmondia trout family. The 96-hour study reported an LC_{50} for mortality of 1,050 mg/L. The most sensitive freshwater invertebrate was the fairy shrimp (*Streptocephalus proboscideus*). The 24-hour study reported an LC_{50} of 1,200 μg/L. The saltwater species were less sensitive than the freshwater species; the single fish record (*Alburnus alburnus*, 96-hour) and the single invertebrate record (*Nitocra spinipes*, 96-hour) reported LC_{50}s of 9,300 mg/L and 4,800 mg/L, respectively.

8.12 HAA: Monobromoacetic Acid

8.12.1 Mammalian Toxicity

Monobromoacetic acid (and other cogeners) is formed on ozination of fulvic acid in the presence of bromide ion (Bull and Kopfler, 1991). The halogenated acetic acids cause severe damage to the skin and mucous membranes that contact these substances. ATSDR has not published a toxicological profile on monobromoacetic acid (ATSDR, 2001). Monobromoacetic acid has not been evaluated by either IARC (2001) or ACGIH (2000).

8.12.1.1 Noncancer Effects

In general, the brominated haloacetic acids are toxic by ingestion, inhalation, and dermal routes (Lewis, 1993). Very few studies are available concerning the toxicology of the brominated haloacetic acids (WHO, 2000). In contrast with dibromoacetic acid, Linder et al. (1994) reported that monobromoacetic acid did not adversely affect reproductive function in male rats at doses as high as 100 mg/kg (single dose) or 25 mg/kg-day (over 14 days). Nevertheless, Linder et al. (1994) reported that the LD_{50} of monobromoacetic acid was nearly ten-fold lower than that of dibromoacetic acid (177 mg/kg vs. 1737 mg/kg, respectively).

8.12.1.2 Carcinogenic Effects

The carcinogenic potential of the brominated haloacetic acids has not been reported (WHO, 2000). The only data available come from *in vitro* mutation assays in bacterial systems by Giller et al. (1997). Giller et al., reported that monobromoacetic acid was negative in two tests for genotoxicity (SOS chromatest at doses up to 1 mg/mL and in the newt micronucleus test). By contrast, Giller found that monobromoacetic acid (in the presence of activated S9) induced bacterial mutation at concentrations as low as 20 μg/mL in the Ames test.

8.12.2 Aquatic Ecotoxicity

A total of 11 unique records were found in AQUIRE. Ten records were excluded from the analysis due to insufficient or incompatible information, i.e., no study endpoint. The remaining record was grouped as a freshwater invertebrate.

The most sensitive acute freshwater effects were seen in an invertebrate, the water flea (*Daphnia magna*). The 24-hour study reported an EC_{50} for immobility of 65,000 ppb. None of the records contained information on freshwater fish or chronic endpoints. None of the records contained information on saltwater fish or invertebrates.

8.13 HAA: Dibromoacetic Acid

Acute exposure to relatively high concentrations of halogenated acetic acids results in severe damage to the skin and mucous membranes; however, these effects are not likely to occur at concentrations seen in treated water. Dibromoacetic acid and other brominated acetic acids are formed in waters that contain bromide, which with the addition of a strong oxidizing agent (like chlorine or ozone) becomes hypobromous acid (WHO, 2000). There are relatively few data on the brominated haloacetic acids with the exception of dibromoacetic acid. Dibromoacetic acid has been identified as a drinking water disinfection by-product under the Information Collection Rule (U.S. EPA, 1994b) and has been well characterized due to its effects on the male reproductive system. ATSDR has not published a toxicological profile on dibromoacetic acid (ATSDR, 2001). Dibromoacetic acid has not been evaluated by either IARC (2001) or ACGIH (2000).

8.13.1 Mammalian Toxicity

No human studies were located for dibromoacetic acid. The overall hazard profile presented by dibromoacetic acid is one of substantial toxicity. This compound had major effects on the male reproductive system resulting in behavioral effects on mating after a relatively short number of exposures and inability to sire litters naturally or by artificial insemination after day 15 of exposure to 250 mg/kg-day. There were more subtle effects on reproductive competence at lower doses including a trend to sire fewer litters with increased dose, which was only statistically significant at 50 mg/kg-day (Linder et al., 1995). There were also severe dose-dependent effects on sperm counts, sperm motility, and sperm morphology down to 10 mg/kg-day (Linder et al., 1995, Linder et al., 1997a). While the mode of action resulting in these effects is not known, it is known that exposure to dichloroacetic acid produces qualitatively similar effects on the male reproductive system.

Dibromoacetic acid also was shown to cause developmental effects. *In vivo* studies showed delayed parturition, effects on prenatal and postnatal pup weights, and kidney damage. In whole embryo culture, a series of 10 haloacetic acids produced many qualitatively similar effects, including effects on neural tube development (Hunter et al., 1996), which were used to derive a quantitative structure-activity relationship largely based on the acid dissociation constant and halogen size (Richard and Hunter, 1996). Developmental effects were generally not seen below 50 mg/kg-day, making them a less sensitive endpoint of toxicity than male reproductive effects. Significant effects on body weight and on absolute and relative liver weight were seen at doses of 250 and 1250 mg/kg-day. Most of the studies reviewed focused on male reproductive effects, and there was limited data on other effects, leaving a somewhat incomplete picture of the toxic effects of dibromoacetic acid.

8.13.1.1 Noncancer Effects

There has been considerable interest in the male reproductive effects of dibromoacetic acid as well as any developmental effects it may have. This area has been the focus of recent studies in part because dichloroacetic acid is known to be a male reproductive toxin (as reported above by Linder et al., 1995, 1997a). An oral study reported on the acute toxicity of dibromoacetic acid as part of a study on its acute spermatogenic effects (Linder et al., 1994). Five male Sprague-Dawley rats per group were dosed with 1000 to 2000 mg/kg dibromoacetic acid (specific dose levels not reported) by gavage in water. Custom-synthesized dibromoacetic acid of >99% purity was used because the commercial chemical is generally contaminated with approximately 10% monobromoacetic acid. Animals were killed 14 -21 days after dosing. The oral LD_{50} was 1737 mg/kg, with a 95% confidence limit range of 1411 to 1952 mg/kg. Animals that died during the observation period generally did so within 48 hours of dosing. Symptoms observed included excess drinking, hypomobility, labored breathing, diarrhea, and ataxia in some animals. The sperm of all surviving animals were degenerating and misshapen, and there was abnormal retention of sperm.

The effects of dibromoacetic acid were studied (Parrish et al., 1996) in the livers of male B6C3F1 mice after exposure to dibromoacetatic acid in drinking water for three weeks at levels of 0.1, 0.5, 2.0 and 3.0 g/L. Six animals per dose group were used. No dose-related change in body weight was observed. Absolute and relative liver weight increased in a dose-dependent manner and reached statistical significance at the two highest doses.

Narotsky et al. (1997) studied developmental effects of dibromoacetic acid in CD-1 mice dosed by gavage with doses equivalent to approximately 0, 24, 50, 100, 200, 392, 610, and 806 mg/kg on gestation days 6 to 15. Maternal effects were limited to piloerection and motor depression at the highest dose. There was delayed parturition at all dose levels. At the highest dose there was increased prenatal mortality with only three of nine litters viable at birth, increased postnatal mortality which was also seen at the 610 mg/kg dose, and decreased pup weight on postnatal day one. Pup weight was also decreased on postnatal day six at the 610 mg/kg dose. Short, kinked, or absent tails were observed at the two highest doses. Dibromoacetic acid was reported to be a clear developmental toxin, and more potent than dichloroacetic acid.

In a more detailed study, dibromoacetic acid was administered to CD-1 mice by gavage in distilled water on gestation days 6-15, at doses of 0, 50, 100, or 400 mg/kg-day (Narotsky et al., 1997). Maternal toxicity was not observed. Litters were removed by cesarean section on gestation day 17, and half of the fetuses in each litter were examined for skeletal defects and the other half for visceral effects. There was no effect on skeletal development. Hydronephrosis was noted at 100 and 400 mg/kg, as well as renal agenesis (small kidneys) at 400 mg/kg. Contrary to the Narotsky et al. (1997) study, which showed delayed parturition at 24 mg/kg, this study showed no developmental effects at the 50 mg/kg dose level, which would represent a NOAEL for developmental effects from this study.

8.13.1.2 Carcinogenic Effects

Under the proposed Guidelines for Carcinogen Risk Assessment (U.S. EPA, 1986), it is concluded that the carcinogenic potential of dibromoacetic acid cannot be determined because no adequate data, such as well-conducted long-term animal studies or reliable human epidemiological studies, are available to perform any assessment. Under the current Risk Assessment Guidelines (U.S. EPA, 1986), dibromoacetic acid is assigned carcinogen category D (not classifiable), indicating inadequate or no human or animal data.

There is, however, suggestive evidence that raises concern for carcinogenic effects. Dichloroacetic acid has been shown to cause liver cancer in experimental animals. Since dichloroacetic and dibromoacetic acids appear to have similar mechanisms of reproductive (Linder et al., 1995; Linder et al., 1997b) and developmental (Hunter et al., 1996) toxicity, it is not unreasonable to anticipate that they may have similar carcinogenic mechanisms. In general, brominated compounds exhibit greater toxicities than their chlorinated analogs.

In vitro assays with dibromoacetic acid showed gene mutations and DNA damage (Giller et al., 1997); *in vivo* assays demonstrated the formation of DNA adducts (Parrish et al., 1996). The only negative genotoxicity assay was the *in vivo* newt micronucleus assay (Giller et al.,

1997). Results of more standard tests, such as the Ames assay using standard test strains or the rodent micronucleus assay, need to be added to this test battery to provide a more complete picture. Nevertheless, a long-term cancer bioassay will likely be necessary to provide a definitive answer unless it can be determined that male reproductive effects are a more sensitive endpoint than cancer for this compound.

8.13.2 Aquatic Ecotoxicity

A total of four unique records were found in AQUIRE. All four records were usable and were grouped as one species of freshwater fish.

The most sensitive acute freshwater effects were seen in a fish, the fathead minnow (*Pimephales promelas*). The 96-hour study reported an LC_{50} for mortality of 69,000 ppb. None of the records contained information on freshwater invertebrates or chronic endpoints. None of the records contained information on saltwater fish or invertebrates.

REFERENCES

ACGIH. 2000. *2000 TLVs and BEIs.* American Conference of Governmental Industrial Hygienists. Cincinnati, OH.

ACGIH. 1991. *Threshold limit values and biological exposure indices for 1990-1991.* American Conference of Governmental Industrial Hygienists. Cincinnati, OH.

American Academy of Microbiology. 2001. *Reevaluation of Microbial Water Quality: Powerful New Tools for Detection and Risk Assessment.* American Academy of Microbiology, Washington, D.C.

ATSDR. 2001. *Internet HazDat-Toxicological Profile Query.* Agency for Toxic Substances and Disease Registry (ATSDR). Online at http://www.atsdr.cdc.gov/gsql/toxprof.script. Last updated 2001 when viewed.

Benjes, H.H. 1976. *Cost Estimating Manual-Combined Sewer Overflow Storage and Treatment. EPA-600/2-76-286 December 1976.* National Environmental Research Center, Office of Research and Development, U.S. Environmental Protection Agency, Cincinnati, OH.

Boner, M.C. and B.G. Turner. 2002. *Demonstration Testing of Multiple Wet Weather Treatment Technologies—Columbus, Georgia Program.* Presented at the Water Environmental Federation Specialty Conference, Urban Wet Weather Pollution Controlling Sewer Overflows and Stormwater Runoff, Quebec City, Canada.

Bull R.J., I.M. Sanchez, M.A. Nelson, J.L. Larson, and A.J. Lansing. 1990. Liver tumor induction in B6C3F1 mice by dichloroacetate and trichloroacetate. Toxicology 63:341-359.

Bull, R.J. and F.C. Kopfler. 1991. *Health Effects of Disinfectants and Disinfection By-Products.* Prepared for the American Water Works Research Foundation. Denver, CO.

Cabelli, V.J., A.P. Dufour, M.A. Levin, L.J. McCabe, and P.W. Haberman. 1979. Relationship of microbial indicators to health effects at marine bathing beaches. Am J Public Health 69: 690-696.

Camp, Dresser, & McKee and Moffa & Associates. 1997. *CSO Disinfection Pilot Study Report Addendum 1.* Camp, Dresser, & McKee, Woodbury, NY.

Cantor, K.P., R. Hoover, P. Hartge, T.J. Mason, D.T. Silverman, R. Altman, D.F. Austin, M.A. Child , C.R. Key, and L.D. Marrett. 1987. Bladder cancer, drinking water source, and tap water consumption: a case-control study. J. Nat Cancer Inst 79(6): 1269-79.

Centers for Disease Control. 1999. Achievements in public health, 1900-1999: Control of infectious diseases. Morb Mortal Weekly Rep 48(29):621. 07/30/1999.

Centers for Disease Control. 2001. CDC Surveillance Summaries Webpage. http://www.cdc.gov/mmwr/sursumpv.html. January 15, 2001 (latest modification when viewed).

Chang, L.W., F.B. Daniel, and A.B. DeAngelo. 1992. Analysis of DNA strand breaks induced in rodent liver in vivo, hepatocytes in primary culture, and a human cell line by chlorinated acetic acids and chlorinated acetaldehydes. Environ Mol Mutagen. 20: 277–288.

Chen, Y.S.R., O.J. Sproul, and A.J. Rubin. 1984. *Inactivation of Naegleria Gruberi Cysts by Chlorine Dioxide*. EPA Grant R808150-02-0, Department of Civil Engineering, Ohio State University.

Cicmanec, J.L., L.W. Condie, and G.R. Olson. 1991. 90-day toxicity study of dichloroacetate in dogs. Fundam Appl Toxicol. 17:376-389.

Daniel, F.B., A.B. DeAngelo, J.A. Stober, G.R. Olson, and N.P. Page. 1992. Hepato-carcinogenicity of chloral hydrate, 2-chloroacetaldehyde and dichloroacetic acid in the male B6C3F1 mouse. Fundam Appl Toxicol., 19: 159–168.

DeAngelo, A.B., F.B. Daniel, L. McMillan, P. Wernsing, and R.E. Savage, Jr. 1989. Species and strain sensitivity to the induction of peroxisome proliferation by chloroacetic acids. Toxicol Appl Pharmacol 101: 285-298.

DeAngelo, A.B, M.H. George, and D.E. House. 1999. Hepatocarcinogenicity in the male B6C3F1 mouse following a life-time exposure to dichloroacetic acid in the drinking water: dose-response determination and modes of action. J Toxicol Environ Health 58:485-507.

DeAngelo, A.B., F.B. Daniel, and B.M. Most. 1996. The carcinogenicity of dichloroacetic acid in the male Fischer 344 rat. Toxicology 114:207-221.

DeAngelo, A.B., F.B Daniel, J.A. Stober, and G.R. Olson. 1991. The carcinogenicity of dichloroacetic acid in the male B6C3F1 mouse. Fundam Appl Toxicol., 16: 337–347

DOH. 1993. *New York State Department of Environmental Conservation and New York State Dept. of Health. Regulating Perchloroethylene (Perc) Emissions From Dry Cleaning Machines: An Economic and Public Health Impact Analysis.* NYSDEC Office of Policy and Program Analysis and Division of Air/NYSDOH Bureau of Toxic Substance Assessment. March 1993.

Dosdall, L.M., L.R. Goodwin, R.J. Casey, and L. Noton. 1997. The effect of ambient concentrations of chlorate on survival of freshwater aquatic invertebrates. Water Qual Res J Canada 32(4): 839-854.

Eschenbrenner, A.B. and E. Miller. 1945. Induction of hepatomas in mice by repeated oral administration of chloroform, with observations on sex differences. J Natl Cancer Inst 5: 251-255.

Giller, S., F. Le Curieux, F. Erb, and D. Marzin. 1997. Comparative genotoxicity of halogenated acetic acids found in drinking water. Mutagenesis 12: 321-328

Haas, C.N. 1986. Wastewater disinfection and infectious disease risks. *CRC Critical Reviews in Environmental Control* 17(1): 1-20.

Herren-Freund, S.L., M.A. Pereira, M.D. Khoury, and G. Olson. 1987. The carcinogenicity of trichloroethylene and its metabolites, trichloroacetic acid and dichloroacetic acid, in mouse liver. Toxicol Appl Pharmacol 90: 183–189.

Heywood, R., R.J. Sortwell, P.R. Noel, A.E. Street, D.E. Prentice, F.J. Roe, P.F. Wadsworth, A.N. Worden, and N.J. Van Abbe. 1979. Safety evaluation of toothpaste containing chloroform. III. Long-term study in beagle dogs. J Environ Pathol Toxicol 2: 835-851.

Hunter, E.S., III, E.H. Rogers, J.E. Schmid, and A. Richard. 1996. Comparative effects of haloacetic acids in whole-embryo culture. Teratology 54: 57–64.

IARC. 2001. *International Agency for Research on Cancer (IARC). Cumulative Cross Index to IARC Monographs on the Evaluation of Carcinogenic Risks to Humans*. IARC Monographs. Online at http://193.51.164.11/cgi/iHound/Chem/iH_Chem_Frames.html.

Jorgenson, T., E. Meierhenry, and C. Rushbrook. 1985. Carcinogenicity of chloroform in drinking water to male Osborne-Mendel rats and female B6C3F1 mice. Fundam. Appl. Toxicol 5(4): 760-769.

Kallen, B.A. and E. Robert. 2000. Drinking water chlorination and delivery outcome—A registry-based study in Sweden. Reprod Toxicol 14(4): 303-9.

Kuhn, R., M. Pattard, K.D. Pernak, and A. Winter. 1989. Results of the harmful effects of water pollutants of Daphnia magna in the 21-day reproduction test. Wat Res 23: 501-510

Kurokawa, Y., S. Takayama, and Y. Konishi. 1986. Long-term in vivo carcinogenicity tests of potassium bromate, sodium hypochlorite, and sodium chlorite conducted in Japan. Environ Health Perspect 69:221-235.

Larson, J. L. and R.J. Bull. 1992. Metabolism and lipoperoxidative activity of trichloroacetate and dichloroacetate in rats and mice. Toxicol Appl Pharmacol 115: 268–277.

Levy, D.A., M.S. Bens, G.F. Craun, R.L Calderon, and B.L. Herwaldt. 1998. Surveillance for waterborne-disease outbreaks—United States, 1995-1996. Morb Mortal Weekly Rep 47/(SS-5): 1-34.

Lewis, R.J., Sr. 1993. *Hawley's Condensed Chemical Dictionary,* 12th Ed. New York: Van Nostrand Reinhold. p.419.

Life Magazine. 2000. The most important events of the millennium. http://www.lifemag.com/Life/millennium/events/100.html. Time Life Magazine. December 29, 2000 (latest modification when viewed).

Linder, R.E., G.R. Klinefelter, L.F. Strader, J.D. Suarez, and C.J. Dyer. 1994. Acute spermatogenic effects of bromoacetic acids. Fundam Appl Toxicol 22: 422–430.

Linder, R.E., G.R. Klinefelter, L.F. Strader, D.N.R. Veeramachaneni, N.L. Roberts, and J.D. Suarez. 1997a. Histopathologic changes in the testes of rats exposed to dibromoacetic acid. Reprod Toxicol 11: 47–56.

Linder, R.E., G.R. Klinefelter, L.F. Strader. 1995. Dibromoacetic acid affects reproductive competence and sperm quality in the male rat. Fundam Appl Toxicol 28: 9–17.

Linder, R.E., G.R. Klinefelter, L.F. Strader, J.D. Suarez, and N. L. Roberts. 1997b. Spermatotoxicity of dichloroacetic acid. Reprod Tox 11:681-688.

Liss, P.S. and P.G. Slater. 1974. Flux of gases across the air-sea interface. Nature 247: 181–184.

Lyman, W.J., W.F. Reehl, D.H. Rosenblatt. 1990. *Handbook of Chemical Property Estimation Methods*. Washington, DC: American Chemical Society. p. 5-4, 5-10.

Mannan, G. and M.K. O'Connor. 2001. *Chemical Incident Data Mining and Application to Chemical Safety Trend Analysis*. Mary Kay O'Connor Process Safety Center, Texas A&M University.

Merck. 1989. *The Merck Index, An Encyclopedia of Chemicals, Drugs, and Biologicals.* 11th ed. Ed. S. Budavari. Rahway, NJ: Merck and Co. Inc.

Missoula County Webpage. 2000. http://www.co.missoula.mt.us/emergencymanagement/hx.html.

Moffa & Associates. 2000. *Newell Street Regional Treatment Facility Disinfection Technology Pilot Demonstration Draft Final Report*. Prepared for the New York State Energy Research and Development Authority and the Onondaga County Department of Drainage and Sanitation. Moffa and Associates Consulting Engineers LLP, Syracuse, NY.

Moore, A.C., B.L. Herwaldt, G.F. Craun, R.L. Calderon, A.K. Highsmith, and D.D. Juranek. 1994. Waterborne disease in the United States, 1991 and 1992. Journal AWWA, Feb. 1994, 87-99.

Moore, G.W., L.L. Swift, D. Rabinowitz, O.B. Crofford, J.A. Oates, and P.W. Stacpoole. 1979. Reduction in serum cholesterol in two patients with homozygous familial hypercholestremia by dichloroacetate. Atherosclerosis 33: 285-293.

Morris, J.C. 1975. *Aspects of the Quantitative Assessment of Germicidal Efficiency in Disinfection: Water and Wastewater*. J.D. Johnson, Editor. Ann Arbor, MI: Ann Arbor Science Publishers, Inc.

Narotsky, M.G., R.A. Pegram, and R.J. Kavlock. 1997. Effect of dosing vehicle on the developmental toxicity of bromodichloromethane and carbon tetrachloride in rats. Fundament Appl Toxicol 40:30-36.

NAS. 1987. *Drinking water and health: disinfectants and disinfection byproducts*. National Academy of Science (NAS). Vol. 7. Washington, D.C.: National Academy Press.

National Response Center. 2001. National Response Center Webpage. http://www.nrc.uscg.mil/nrchp.html. Last modified when viewed 2001.

Nelson, M.A. and R.J. Bull.1988. Induction of strand breaks in DNA by trichloroethylene and metabolites in rat and mouse liver in vivo. Toxicol Appl Pharmacol, 94: 45–54.

NIOSH. 1991. *Registry of Toxic Effects of Chemical Substances: Trichloroacetic Acid.* Cincinnati, OH: U.S. Department of Health and Human Services, Public Health Service, Centers for Disease Control, National Institute for Occupational Safety and Health, Division of Standards Development and Technology Transfer, Technical Information Branch.

NLM. 1992. *Hazardous Substances Data Bank: Trichloroacetic Acid*. Bethesda, MD: National Library of Medicine.

NSF International and U.S. EPA. 2002. Verification of High-Rate Disinfection—Induction Mixers. http://www.epa.gov/etv/verifications/vcenter9-2.html

NTP (National Toxicology Program). 1985. *Toxicology and Carcinogenesis Studies of Chlorodibromomethane in F344/N Rats and B6C3F1 mice (gavage studies)*. NTP TR282.

NTP (National Toxicology Program). 1986. *Toxicology and carcinogenesis studies of bromodichloromethane in F344/N rats and B6C3F mice (gavage studies)*. NTP Technical Report, Ser. No. 321, NIH Publ. No. 87-2537.

NTP (National Toxicology Program). 1988. *Technical Report on the Toxicology and Carcinogenesis Studies of Tribromomethane (Bromoform) (CAS No. 75-25-2) in F344 rats and B6C3F1 mice (gavage studies)*. NTP TR 350.

NTP (National Toxicology Program). 1989. *Toxicology and Carcinogenicity Studies of Tribromomethane and Bromoform in F344/N rats and B6C3F1 Mice (gavage study).* NTP-350. Research Triangle Park, NC.

NTP (National Toxicology Program). 1992. National Toxicology Program (NTP). *NTP Technical Report on the Toxicology and Carcinogenesis Studies of Chlorinated and Chloraminated Water in F344/N Rats and B6C3F1 Mice (drinking water studies).* U.S. Department of Health and Human Services. NTP TR 392.

Ono, Y., I. Somiya, and M. Kawamura. 1991. The evaluation of genotoxicity using DNA repairing test for chemicals produced in chlorination and ozonation processes. Wat Sci Tech 23: 329-338.

Orum, P. 2001. Testimony of Paul Orum Working Group on Community Right-to-Know Before the Subcommittee on Superfund, Toxics, Risk, and Waste Management of the Senate Environment and Public Works Committee. November 24, 2001.

Parrish, J.M., E.W. Austin, and D.K. Stevens. 1996. Haloacetate-induced oxidative damage to DNA in the liver of male B6C3F1 mice. Toxicology 110:103-111.

Pelczar, M.J., Jr, E.C.S. Chan, and N.R. Krieg. 1993. *Microbiology: Concepts and Applications.* New York: McGraw-Hill.

Pereira, M.A. 1983. Carcinogenicity of chlorination by-products: trihalomethanes. In: *Water Chlorination: Environmental Impact and Health Effects, Vol. 4. Environment, Health, and Risk, Book 2*, R.L. Jolley, W.A. Brungs, J.A. Cotruvo, R.B. Cumming, J.S. Mattice, V.A. Jacobs, Ed. Ann Arbor Science Publishers, Inc., Ann Arbor, MI. p. 1165-1176.

Pereira, MA. 1996. Carcinogenic activity of dichloroacetic acid and trichloroacetic acid in the liver of female B6C3F1 mice. Fundam Appl Toxicol 31:192-199.

Pereira, M.A. and J.B. Phelps. 1996. Promotion by dichloroacetic acid and trichloroacetic acid of N-methyl-N-nitrosourea-initiated cancer in the liver of female B6C3F1 mice. Cancer Lett 102:133-141.

Richard, A.M. and E.M Hunter. 1996. Quantitative structure-activity relationships for the developmental toxicity of haloacetic acids in mammalian whole embryo culture. Teratology 53:352-360.

Richmond, R.E., J.H. Carter, H.W. Carter, F.B. Daniel, and A.B. DeAngelo. 1995. Immunohistochemical analysis of dichloroacetic acid (DCA)-induced hepatocarcinogenesis in male Fischer (F344) rats. Cancer Lett 92:67-76.

Rosenberg, M.L., J.P. Koplan, I.K. Wach-smuth, B. Cline, W.E. DeWitt, P. Skaliy, and G.F. Mallison. 1975. *Transmission of Shigellosis by Swimming in a Contaminated River*. Public Health Service, Centers for Disease Control, Atlanta, GA.

Sandman, P.M. 1993. *Responding to Community Outrage: Strategies for Effective Risk Communication.* A publication of the American Industrial Hygiene Association.

Sandman, P.M. 2003. The Peter Sandman Risk Communication Website. http://www.psandman.com/.

Stacpoole, P.W. 1990. The pharmacology of dichloroacetate. Metabolism 38, 1124-1144.

Stinson, M.K., J.S. Wojtenko, and R. Field. 1999. U.S. EPA in-house survey of CSO treatment facilities in the United States. Edison, NJ: U.S. EPA.

Symons, J.M. 2001. The early history of disinfection by-products: a personal chronicle (part 1). Environmental Engineer. 37 (1): 20-26. 2001.

The Royal Society of Chemistry. 1983. *The Agrochemicals Handbook.* Nottingham, England: The University.

U.S. EPA. 1975. *Bench-Scale High Rate Disinfection of Combined Sewer Overflows with Chlorine and Chlorine Dioxide*. Office of Research and Development, National Environmental Research Center, Cincinnati, OH. EPA-670/2-75-021.

U.S. EPA. 1979a. *Disinfection/ Treatment of Combined Sewer Overflows, Syracuse, NY.* National Environmental Research Center, Office of Research and Development, U.S. Environmental Protection Agency, Cincinnati, OH 45268. EPA-600/2-79-134.

U.S. EPA. 1979b. *Combined Sewer Overflow Abatement Program-Rochester, NY Volume II.* Pilot Plant Evaluations. Office of Research and Development, National Environmental Research Center, Cincinnati, OH. EPA-600/2-79-031b.

U.S. EPA. 1985. *EPA Chemical Profile for Chloroacetic Acid. EPCRA Information Specialist*. Washington, D.C. 1985.

U.S. EPA. 1986. *Guidelines for Carcinogen Risk Assessment*. 51 Federal Register 33992-34003. September 24, 1986.

U.S. EPA. 1989. *Risk Assessment Guidance for Superfund, Volume I, Human Health Evaluation Manual (Part A)*. U.S. Environmental Protection Agency Washington, D.C. Interim Final EPA/540/1-89/002.

U.S. EPA. 1991. *CARA (Chemical Assessments and Related Activities). April, 1991*. Office of Health and Environmental Assessment. Cincinnati, OH.

U.S. EPA. 1992a. *Framework for Ecological Risk Assessment*. Washington, D.C.: Risk Assessment Forum; EPA/630/R-92/001.

U.S. EPA. 1992b. *Drinking Water Criteria Document for Chlorine, Hypochlorous Acid and Hypochlorite Ion*. Prepared by the Office of Health and Environmental Assessment, Environmental Criteria and Assessment Office, Cincinnati, OH for the Office of Water, Washington, D.C.

U.S. EPA. 1994a. *CARA (Chemical Assessments and Related Activities). December, 1994*. Office of Health and Environmental Assessment. Cincinnati, OH.

U.S. EPA. 1994b. *Technical Background Document to Support Rulemaking Pursuant to the Clean Air Act C Section 112(g). Ranking of Pollutants with Respect to Hazard to Human Health. EPAB450/3-92-010*. Emissions Standards Division, Office of Air Quality Planning and Standards, Research Triangle Park, NC. 1994.

U.S. EPA. 1997a. *Health Effects Assessment Summary Tables. FY-1997 Update*. Office of Research and Development, Office of Emergency and Remedial Response, Washington, D.C. EPA/540/R-97/036. NTIS PB 97-921199.

U.S. EPA. 1997b. *Ecological Risk Assessment Guidance for Superfund: Process for Designing and Conducting Ecological Risk Assessments—Interim Final*. U.S. EPA Office of Solid Waste and Emergency Response, Washington, D.C. June 1997. EPA 540-R-97-006.

U.S. EPA. 1998. *National Primary Drinking Water Regulations: Disinfectants and Disinfection By-Products; Notice of Data Availability; Final Rule. Fed Reg 63:69390-476*. December 16. Stage 1 Disinfectants and Disinfection By-Products Rule, EPA 815-F-98-010.

U.S. EPA. 1999a. *Combined Sewer Overflow Technology Fact Sheet Alternative Disinfection Methods*. Washington, D.C.: U.S. Environmental Protection Agency. EPA 832-F-99-033.

U.S. EPA. 1999b. Combined *Sewer Overflow Technology Fact Sheet Chlorine Disinfection*. Washington, D.C.: U.S. Environmental Protection Agency. EPA 832-F-99-034.

U.S. EPA. 1999c. *Alternative Disinfectants and Oxidants Guidance Manual*. Washington, D.C.: U.S. Environmental Protection Agency. EPA 815-R-99-014.

U.S. EPA. 1999d. *Guidance for conducting health risk assessment of chemical mixtures (External scientific peer review draft)*. Washington, D.C.: U.S. Environmental Protection Agency, Risk Assessment Forum. NCEA-C-0148. Online at http://www.epa.gov/NCEA/mixtures.htm.

U.S. EPA. 2000a. *About the NPDES Storm Water Program*. http://www.epa.gov/owm/sw/about/. U.S. Environmental Protection Agency, Office of Water Management, Washington, D.C., June 05, 2000 (latest modification when viewed).

U.S. EPA. 2000b. CSO Homepage. http://www.epa.gov/owm/cso.htm. U.S. Environmental Protection Agency, Office of Water Management, Washington, D.C. October 19, 2000 (latest modification when viewed).

U.S. EPA. 2000c. SSO Homepage. http://www.epa.gov/owm/sso.htm. U.S. Environmental Protection Agency, Office of Water Management, Washington, D.C. April 12, 2000 (latest modification when viewed).

U.S. EPA. 2000d. Toxicological review of chlorine dioxide and chlorite. In Support of Summary Information on the Integrated Risk Information System (IRIS). September 2000. EPA/635/R-00/007.

U.S. EPA. 2001. *Integrated Risk Information System (IRIS)*. U.S. EPA Office of Research and Development. National Center or Environmental Assessment, Washington, D.C. Online at http://www.epa.gov/iris/.

U.S. EPA. 2002. *Implementation Guidance for Ambient Water Quality Criteria for Bacteria May 2002 Draft*. U.S. Environmental Protection Agency Office of Water Washington, D.C. EPA-823-B-02-003. U.S. EPA. 2003. US *EPA Chlorine Hazard Summary Webpage*. http://www.epa.gov/ttn/atw/hlthef/chlorine.html. Last updated on 2/12/03 when viewed.

U.S. EPA. 2004a. *Risk Assessment Guidance for Superfund (RAGS), Volume I: Human Health Evaluation Manual (Part E, Supplemental Guidance for Dermal Risk Assessment)*. Office of Superfund Remediation and Technology Innovation U.S. Environmental Protection Agency Washington, D.C. EPA/540/R/99/005.

U.S. EPA. 2004b. *Integrated Risk Information System (IRIS)*. U.S. EPA Office of Research and Development. National Center for Environmental Assessment, Washington, D.C. Online at http://www.epa.gov/iris/.

Water Environment Federation. 1996. *Wastewater Disinfection Manual of Practice No. FD-10.* Alexandria, VA: Water Environment Federation.

Water Pollution Control Federation. 1984. *Wastewater Disinfection, A State-of-the-Art Report.* Washington, D.C.: Water Pollution Control Federation. p. 78.

Werdehoff, K.S. and P.C. Singer. 1987. Chlorine dioxide effects on THMFP, TOXFP and the formation of inorganic by-products. Journal AWWA. 79(9):107.

White, G.C. 1999. *Handbook of Chlorination and Alternative Disinfection.* 4th Edition. New York: Van Nostrand Reinhold.

WHO. World Health Organization (WHO). 2000. *Disinfectants and Disinfectant By-Products. International Programme on Chemical Safety, Environmental Health Criteria Document No. 216*. Geneva: WHO.

WHO. World Health Organization (WHO). 2001. *WHO Online Library Catalog*. Online at http://chef.who.int:9654/?WHOhq+WHOhqHTML.

Yokose, Y; Uchida, K; Nakae, Dl. 1987. Studies of carcinogenicity of sodium chlorite in B6C3F1 mice. Environ Health Perspect 76:205-210.

WERF SUBSCRIBERS

WASTEWATER UTILITY

Alabama
Montgomery Water Works & Sanitary Sewer Board

Alaska
Anchorage Water & Wastewater Utility

Arizona
Gila Resources
Glendale, City of, Utilities Department
Mesa, City of
Peoria, City of
Phoenix, City of
Pima County Wastewater Management

Arkansas
Little Rock Wastewater Utility

California
Calaveras County Water District
Central Contra Costa Sanitary District
Corona, City of
Crestline Sanitation District
Delta Diablo Sanitation District
Dublin San Ramon Services District
East Bay Dischargers Authority
East Bay Municipal Utility District
Eastern Municipal Water District
El Dorado Irrigation District
Fairfield-Suisun Sewer District
Fresno Department of Public Utilities
Inland Empire Utilities Agency
Irvine Ranch Water District
Las Virgenes Municipal Water District
Livermore, City of
Lodi, City of
Los Angeles, City of
Los Angeles County, Sanitation Districts of
Napa Sanitation District
Orange County Sanitation District
Palo Alto, City of
Riverside, City of
Sacramento Regional County Sanitation District
San Diego Metropolitan Wastewater Department, City of
San Francisco, City & County of
San Jose, City of
Santa Barbara, City of
Santa Cruz, City of
Santa Rosa, City of
South Bayside System Authority
South Coast Water District
South Orange County Wastewater Authority
Stege Sanitary District
Sunnyvale, City of
Union Sanitary District
West Valley Sanitation District

Colorado
Aurora, City of
Boulder, City of
Colorado Springs Utilities
Greeley, City of
Littleton/Englewood Water Pollution Control Plant
Metro Wastewater Reclamation District, Denver

Connecticut
The Mattabassett District
New Haven, City of, WPCA

District of Columbia
District of Columbia Water & Sewer Authority

Florida
Broward, County of
Fort Lauderdale, City of
JEA
Miami-Dade Water & Sewer Authority
Orange County Utilities Department
Reedy Creek Improvement District
Seminole County Environmental Services
St. Petersburg, City of
Stuart Public Utilities
Tallahassee, City of
Tampa, City of
Toho Water Authority
West Palm Beach, City of

Georgia
Atlanta Department of Watershed Management
Augusta, City of
Clayton County Water Authority
Cobb County Water System
Columbus Water Works
Fulton County
Gwinnett County Department of Public Utilities
Savannah, City of

Hawaii
Honolulu, City & County of

Idaho
Boise, City of

Illinois
American Bottoms Wastewater Treatment Plant
Greater Peoria Sanitary District
Kankakee River Metropolitan Agency
Metropolitan Water Reclamation District of Greater Chicago
Wheaton Sanitary District

Iowa
Ames, City of
Cedar Rapids Wastewater Facility
Des Moines Metro Wastewater Reclamation Authority
Iowa City

Kansas
Johnson County Unified Wastewater Districts
Unified Government of Wyandotte County/ Kansas City, City of

Kentucky
Louisville & Jefferson County Metropolitan Sewer District

Louisiana
Sewerage & Water Board of New Orleans

Maine
Bangor, City of
Portland Water District

Maryland
Anne Arundel County Bureau of Utility Operations
Hagerstown, City of
Howard County Department of Public Works
Washington Suburban Sanitary Commission

Massachusetts
Boston Water & Sewer Commission
Upper Blackstone Water Pollution Abatement District

Michigan
Ann Arbor, City of
Detroit, City of
Holland Board of Public Works
Lansing, City of
Saginaw, City of
Wayne County Department of Environment
Wyoming, City of

Minnesota
Rochester, City of
Western Lake Superior Sanitary District

Missouri
Independence, City of
Kansas City Missouri Water Services Department
Little Blue Valley Sewer District
Metropolitan St. Louis Sewer District

Nebraska
Lincoln Wastewater System

Nevada
Henderson, City of
Reno, City of

New Jersey
Bergen County Utilities Authority
Ocean, County of
Passaic Valley Sewerage Commissioners

New York
New York City Department of Environmental Protection

North Carolina
Charlotte/Mecklenburg Utilities
Durham, City of
Metropolitan Sewerage District of Buncombe County
Orange Water & Sewer Authority

Ohio
Akron, City of
Butler County Department of Environmental Services
Columbus, City of
Metropolitan Sewer District of Greater Cincinnati
Northeast Ohio Regional Sewer District
Summit, County of

Oklahoma
Tulsa, City of

Oregon
Clean Water Services
Eugene/Springfield Water Pollution Control
Water Environment Services

Pennsylvania
Philadelphia, City of
University Area Joint Authority

South Carolina
Charleston Commissioners of Public Works
Mount Pleasant Waterworks & Sewer Commission
Spartanburg Sanitary Sewer District

Tennessee
Cleveland, City of
Knoxville Utilities Board
Murfreesboro Water & Sewer Department
Nashville Metro Water Services

Texas
Austin, City of
Dallas Water Utilities
Denton, City of
El Paso Water Utilities
Fort Worth, City of
Gulf Coast Waste Disposal Authority
Houston, City of
San Antonio Water System
Trinity River Authority

Utah
Salt Lake City Corporation

WERF SUBSCRIBERS

Virginia
Alexandria Sanitation Authority
Arlington, County of
Fairfax County
Hampton Roads Sanitation District
Henrico, County of
Hopewell Regional Wastewater Treatment Facility
Loudoun County Sanitation Authority
Lynchburg Regional WWTP
Prince William County Service Authority
Richmond, City of
Rivanna Water & Sewer Authority

Washington
Edmonds, City of
Everett, City of
King County Department of Natural Resources
Seattle Public Utilities
Sunnyside, Port of
Yakima, City of

Wisconsin
Green Bay Metro Sewerage District
Kenosha Water Utility
Madison Metropolitan Sewerage District
Milwaukee Metropolitan Sewerage District
Racine, City of
Sheboygan Regional Wastewater Treatment
Wausau Water Works

Australia
South Australian Water Corporation
Sydney Water Corporation
Water Corporation of Western Australia

Canada
Greater Vancouver Regional District
Regina, City of, Saskatchewan
Toronto, City of, Ontario
Winnipeg, City of, Manitoba

Mexico
Servicios de Agua y Drenaje de Monterrey, I.P.D.

New Zealand
Watercare Services Limited

United Kingdom
Yorkshire Water Services Limited

STORMWATER UTILITY

California
Los Angeles, City of, Department of Public Works
Monterey, City of
San Francisco, City & County of
Santa Rosa, City of

Colorado
Aurora, City of
Boulder, City of

Georgia
Griffin, City of

Iowa
Cedar Rapids Wastewater Facility
Des Moines Metro Wastewater Reclamation Authority

Kansas
Overland Park, City of

Maine
Portland Water District

Minnesota
Western Lake Superior Sanitary District

North Carolina
Charlotte, City of, Stormwater Services

Pennsylvania
Philadelphia, City of

Tennessee
Chattanooga Stormwater Management

Washington
Bellevue Utilities Department
Seattle Public Utilities

STATE

Arkansas Department of Environmental Quality
Fresno Metropolitan Flood Control District
Kansas Department of Health & Environment
Kentucky Department of Environmental Protection
Ohio River Valley Sanitation Commission
Urban Drainage & Flood Control District, CO

CORPORATE

Abt. Associates
ADS Environmental Services
The ADVENT Group Inc.
Alan Plummer & Associates
Alden Research Laboratory Inc.
American Water
Aqua-Aerobic Systems Inc.
Aquateam–Norwegian Water Technology Centre A/S
ARCADIS
Associated Engineering
Black & Veatch
Blasland, Bouck & Lee Inc.
Boyle Engineering Corporation
Brown & Caldwell
Burns & McDonnell
CABE Associates Inc.
The Cadmus Group
Camp Dresser & McKee Inc.
Carollo Engineers Inc.
Carpenter Environmental Associates Inc.
CDS Technologies Inc.
Chemtrac Systems Inc.
CH2M HILL
Construction & Environmental Consulting, Inc.
Damon S. Williams Associates, LLC
David L. Sheridan, P.C.
Dewling Associates Inc.
Earth Tech Inc.
Ecovation
EMA Inc.
The Eshelman Company Inc.
EWT Holdings Corporation
Fay, Spofford, & Thorndike Inc.
Freese & Nichols Inc.
ftn Associates Inc.
Fuss & O'Neill Inc.
Gannett Fleming Inc.
Geosyntec Consultants
GHD
Greeley and Hansen LLC
Hazen & Sawyer, P.C.
HDR Engineering Inc.
HNTB Corporation
HydroQual Inc.
Infilco Degremont Inc.
Institute for Environmental Technology & Industry, Korea
Jacobson Helgoth Consultants Inc.
Jason Consultants LLC Inc.
Jordan, Jones, & Goulding Inc.
KCI Technologies Inc.
Kelly & Weaver, P.C.
Kennedy/Jenks Consultants
KMK Consultants
Komline Sanderson Engineering Corporation
Lawler, Matusky & Skelly Engineers, LLP
Limno-Tech Inc.
LTI Inc.
Lombardo Associates Inc.
Malcolm Pirnie Inc.
Material Matters
McKim & Creed
MEC Analytical Systems Inc.
Metcalf & Eddy Inc.
MWH
New England Organics
O'Brien & Gere Engineers Inc.
Odor & Corrosion Technology Consultants Inc.
Oswald Green, LLC
PA Government Services Inc.
Parametrix Inc.
Parsons
Post, Buckley, Schuh & Jernigan
R&D Engineering/Conestoga Rover & Associates
RMC Inc.
R.M. Towill Corporation
Ross & Associates Ltd.
Rothberg, Tamburini & Windsor, Inc.
Royce Technologies
SAIC Maritime Technical Group
Stantec Consulting Inc.
Stearns & Wheler, LLC
Stone Environmental Inc.
Stormwater360
Stratus Consulting Inc.
Synagro Technologies Inc.
Tetra Tech Inc.
Trojan Technologies Inc.
URS Corporation
USfilter
Wade-Trim Inc.
Westin Engineering Inc.
Weston Solutions Inc.
Woodard & Curran
WRc/D&B, LLC
WWETCO, LLC
Zenon Environmental Inc.
Zoeller Pump Company

INDUSTRY

American Electric Power
ChevronTexaco Energy Research & Technology Company
The Coca-Cola Company
Dow Chemical Company
DuPont Company
Eastman Chemical Company
Eastman Kodak Company
Eli Lilly & Company
Merck & Company Inc.
Procter & Gamble Company
RWE Thames Water Plc
Severn Trent Services Inc.
Suez Environnment
United Water Services LLC

Note: List as of 10/1/05

WERF Product Order Form

a benefit of joining the Water Environment Research Foundation, subscribers are entitled to receive one complimentary copy of all final orts and other products. Additional copies are available at cost (usually $10). To order your complimentary copy of a report, please write e" in the unit price column. WERF keeps track of all orders. If the charge differs from what is shown here, we will call to confirm the total ore processing.

ne ______________________ Title ______________________

anization ______________________

ress ______________________

______ State ______ Zip Code ______ Country ______

ne ______ Fax ______ Email ______

Stock #	Product	Quantity	Unit Price	Total
			Postage & Handling	
			VA Residents Add 4.5% Sales Tax	
			Canadian Residents Add 7% GST	
			TOTAL	

Method of Payment:

(All orders must be prepaid.)

Check or Money Order Enclosed

Visa ❑ Mastercard ❑ American Express

ount No. ______________________ Exp. Date ______

nature ______________________

Shipping & Handling:

Amount of Order	United States	Canada & Mexico	All Others
Up to but not more than:	Add:	Add:	Add:
$20.00	$5.00*	$8.00	50% of amount
30.00	5.50	8.00	40% of amount
40.00	6.00	8.00	
50.00	6.50	14.00	
60.00	7.00	14.00	
80.00	8.00	14.00	
100.00	10.00	21.00	
150.00	12.50	28.00	
200.00	15.00	35.00	
More than $200.00	Add 20% of order	Add 20% of order	

*minimum amount for all orders

e: Please make checks payable to the Water Environment Research Foundation.

To Order (Subscribers Only):

Log on to www.werf.org and click on the "Product Catalog."

Phone: (703) 684-2470
Fax: (703) 299-0742.

WERF
Attn: Subscriber Services
635 Slaters Lane
Alexandria, VA 22314-1177

To Order (Non-Subscribers):

Non-subscribers may be able to order WERF publications either through WEF (www.wef.org) or IWAP (www.iwapublishing.com). Visit WERF's website at www.werf.org for details.

IWA Publishing's authorised EU representative for General Product Safety Regulations is Diane D'Arras, 15 rue Duret, 75116 Paris, France, e-mail: safety@iwap.co.uk.

Printed and bound by CPI Group (UK) Ltd, Croydon, CR0 4YY
12/05/2026
02108868-0015